Multimodal Imaging in Neurology: Special focus on MRI applications and MEG

Multimodal Imaging in Neurology: Special focus on MRI applications and MEG

Hans-Peter Müller and Dr. Jan Kassubek

ISBN: 978-3-031-00495-7   paperback
ISBN: 978-3-031-01623-3   ebook

DOI 10.1007/978-3-031-01623-3

A Publication in the Springer series

SYNTHESIS LECTURES ON BIOMEDICAL ENGINEERING #16

Lecture #16
Series Editor : John D. Enderle, University of Connecticut

Series ISSN
ISSN 1930-0328   Print
ISSN 1930-0336   Electronic

Multimodal Imaging in Neurology: Special focus on MRI applications and MEG

Hans-Peter Müller and Jan Kassubek
University of Ulm, Germany

SYNTHESIS LECTURES ON BIOMEDICAL ENGINEERING #16

ABSTRACT

The field of brain imaging is developing at a rapid pace and has greatly advanced the areas of cognitive and clinical neuroscience. The availability of neuroimaging techniques, especially magnetic resonance imaging (MRI), functional MRI (fMRI), diffusion tensor imaging (DTI) and magnetoencephalography (MEG) and magnetic source imaging (MSI) has brought about breakthroughs in neuroscience.

To obtain comprehensive information about the activity of the human brain, different analytical approaches should be complemented. Thus, in "intermodal multimodality" imaging, great efforts have been made to combine the highest spatial resolution (MRI, fMRI) with the best temporal resolution (MEG or EEG). "Intramodal multimodality" imaging combines various functional MRI techniques (e.g., fMRI, DTI, and/or morphometric/volumetric analysis). The multimodal approach is conceptually based on the combination of different noninvasive functional neuroimaging tools, their registration and cointegration. In particular, the combination of imaging applications that map different functional systems is useful, such as fMRI as a technique for the localization of cortical function and DTI as a technique for mapping of white matter fiber bundles/tracts.

This booklet gives an insight into the wide field of multimodal imaging with respect to concepts, data acquisition, and postprocessing. Examples for intermodal and intramodal multimodality imaging are also demonstrated.

KEYWORDS

diffusion tensor imaging, functional magnetic resonance imaging, intermodal multimodality, intramodal multimodality, magnetic resonance imaging, magnetoencephalography, multimodal imaging

Preface

The field of brain imaging is developing at a rapid pace and has greatly advanced the areas of cognitive and clinical neuroscience. Cognitive neuroscience aims at understanding the functional and neural architecture of cognitive functions. The availability of neuroimaging techniques, especially magnetic resonance imaging (MRI), functional MRI (fMRI), diffusion tensor imaging (DTI), and magnetoencephalography (MEG)/magnetic source imaging (MSI) has brought many new insights in cognitive neuroscience as well as in the understanding, and diagnosis of brain diseases.

To obtain comprehensive information about the activity of the human brain, various analytical approaches and different methods of activation should be complemented. Thus, in "intermodal multimodality" imaging, great efforts have been performed to combine the highest spatial resolution (MRI, fMRI) with the best temporal resolution (MEG or EEG). Additionally, "intramodal multimodality" imaging combines various functional MRI techniques (DTI, fMRI, and/or morphometry). Comprehensive information can be provided by radionuclide techniques (PET or SPECT) or by MR spectroscopy (MRS).

The multimodal approach is conceptually based on the combination of different noninvasive functional neuroimaging tools, their registration and cointegration. In particular, intramodal multimodality is useful, for example, the combination of imaging applications on the same technical approach, such as MRI that maps different functional systems, fMRI as a technique for the localization of cortical function, and DTI as a technique for mapping of white matter fiber bundles/tracts. MSI as an additional imaging technique of cortical function is of particular advantage in epilepsy diagnostics.

This booklet provides first insights for engineers or medical doctors who deal with multimodal imaging. Basic measurement techniques as well as procedures for the overlay of results from different measurement techniques with a special focus on MRI-based techniques and MSI are described in detail. Examples for intermodal and intramodal multimodality imaging are demonstrated. The authors have tried to do away with the basic details and formulas where possible, nevertheless have given formulas and mathematics where the understanding is assisted.

Acknowledgments

Dr. Alexander Unrath, University of Ulm, Germany, is thankfully acknowledged for his engagement in course of the DTI experiments and intramodal integration.

Dr. Sergio Nicola Erné, Friedrich Schiller University, Jena, Germany, is thankfully acknowledged for his engagement in course of the MEG experiments and the intermodal integration.

Prof. Dr. Siegfried Stapf, Technical University Illmenau, Germany, is thankfully acknowledged for assistance in the MRI chapters.

Contents

CHAPTER 1

INTRODUCTION

At the beginning of the 20th century, neuroimaging began with a technique called pneumoencephalography [1]. This process basically consisted of draining the cerebrospinal fluid and replacing it with air, altering the relative density of the brain and its surroundings. In this way, X-ray contrasts were improved. Computed tomography (CT) and magnetic resonance imaging (MRI) were discovered between 1970 and 1980. These technologies were considerably less harmful. The next neuroimaging techniques to be developed were single photon emission computed tomography (SPECT) and positron emission tomography (PET) that allowed the mapping of the brain functionalities providing more than just static images of the brain's structure. The end of the last century showed the great breakthrough in mapping human brain functionalities by developing functional MRI (fMRI) with abilities that opened the door to direct observation of cognitive activities. This technique includes, among other options, the study of emotions, social functions (social neuroscience), decision-making (neuroeconomics) and executive functions. Clinical neuroscience aims at understanding, diagnosing and treating diseases of the central nervous system such as neurodegenerative diseases, stroke, epilepsy, and psychiatric disorders. In today's aging society, brain-related diseases are an increasing cost factor. Therefore, both from a scientific and a social point of view, brain imaging is one of the most challenging research areas of the 21st century.

In addition to these localization techniques, the temporal evolution of brain signals may be of interest too. Techniques with a high temporal resolution are electroencephalography (EEG) and magnetoencephalography (MEG). Both techniques are completely noninvasive. The technical equipment needed for EEG is in principle restricted to electrodes, an amplifier, and a detection device, whereas MEG is a much more complicated technique (for details refer to Section 2.2). The high temporal resolution of EEG/MEG can complement the spatial resolution of CT/MRT for comprehensive information – called intermodal multimodal imaging.

Multimodality imaging is widely considered to involve the incorporation of two or more imaging modalities. The localization of signals detected by MEG or EEG, PET or SPECT has an anatomical background with limited spatial resolution. The combination of these modalities with more morphologically oriented ones such as MRI (or CT) offers great opportunities. These methodologies constrain and complement each other. Combined, they can therefore improve our interpretation of functional neural organization. Therefore, it is useful to investigate the brain functions in a multimodal noninvasive way by combining different neuroimaging techniques, thus making use of their individual advantages. Furthermore, the acquisition of individual volume-rendering three-dimensional (3-D) MRI data allows improved models of the anatomical basis or of the head volume conductor (see Chapter 2). For example, one approach to analyze MEG data with particular respect to the inverse problem is to impose anatomic and physiologic constraints for magnetic source imaging (MSI) derived from other imaging modalities such as fMRI in order to reduce the solution space. Combinations of MEG and fMRI in the same paradigm were described in several studies [2,3,4,5], both by comparing the localization of the results as well as applying fMRI constraints to MEG data analysis. Efforts have also been made for the combination of electroencephalography and fMRI [6]. Another field of research is magnetic resonance spectroscopy (MRS) and its intramodality multimodal integration [7].

The main topic of the following chapters is the multimodal combination presented in the form of short introductions to the various measurement techniques. The focus is on the principles of multimodal imaging. Basically, two ways of multimodal imaging can be distinguished. The first way is intermodal multimodality – results of different measurement techniques have to be combined and transformed to a common coordinate frame. To perform this task modality specific markers have to be used to obtain reference points in each specific modality/measurement technique. As an example the combination of MRI, fMRI, and MEG is explained in detail in Section 4.1.

The second way is called intramodal multimodality. Here, results obtained from one single modality such as MRI are combined in a common coordinate frame. Intersubject comparison, group averaging, and comparison between subject groups with different clinical categorizations can be performed by using, for example, the Montreal Neurological Institute (MNI) coordinate frame as a common basis. Transformation rules have to be set up that allows the transformation of results of any single measurement into this coordinate frame. The focus in Section 4.2 is the comprehensive information of diffusion tensor imaging (DTI) results along with morphometry results, while fMRI results complements the data.

Examples of clinical applications and an outlook of future applications of multimodal imaging will complete the walk through different aspects of multimodal imaging.

CHAPTER 2

Neurological Measurement Techniques and First Steps of Postprocessing

2.1 MAGNETIC RESONANCE IMAGING (MRI)

2.1.1 Principles of Magnetic Resonance

MRI was introduced in clinical routine in the early 1980s. It has become a preferred neuroimaging modality because of its feasibility to generate images of slices in arbitrary orientation with different tissue contrast [8,9,10,11]. Nuclei [protons are preferred as $I = 1/2$, and they possess the largest gyromagnetic ratio (γ, see below)] possess an angular momentum. This is coupled with a magnetic dipole moment, which is proportional to the angular momentum. In the presence of a strong magnetic field a nucleus with quantum number I has $2I + 1$ discrete energy states

$$E_m = -\gamma \hbar B_z m \qquad (2.1)$$

where γ is the gyromagnetic ratio, $\hbar$ is the Planck's constant, B_z is the flux density of magnetic field, m is the magnetic quantum number. A change of energy levels means emission or absorption of a photon with

$$\hbar\omega = E_{m-1} - E_m = \gamma \hbar B_z \quad m = \pm 1 \qquad (2.2)$$

where ω is the angular frequency.

In an external magnetic field, the occupation of energy states [Eq. (2.1)] is in accordance to Boltzmann's statistics

$$\frac{N_{m-1}}{N_m} = e^{-\hbar\omega/kT} \qquad (2.3)$$

where k is the Boltzmann's constant, and T is the temperature.

In this way, a surplus of nuclear spins leads to a macroscopic magnetization align parallel to the magnetic field resulting in an angular momentum and an associated macroscopic magnetic moment. If this parallel alignment is disturbed, a torque acts on the magnetic moment of the sample, resulting in a precession of the magnetization with the Larmor frequency

$$\omega = \gamma B_z \qquad (2.4)$$

In MR, precession of the nuclear magnetization is stimulated by disturbing the alignment of the nuclear magnetization parallel to the main magnetic field by a radio frequency (RF) field according to Eq. (2.4). This is realized by a coil around the sample with its field axis orthogonal to the static magnetic field.

The RF magnetic flux density B_1 of a pulse of duration t leads to a rotation of the macroscopic magnetization about an angle α between the macroscopic magnetization and the static magnetic field

$$\alpha = \gamma B_1 t \tag{2.5}$$

resulting in a precessing nuclear magnetization $\vec{M}$ with components

$$
\begin{aligned}
M_x &= -M_0 \, \sin(\alpha) \, \sin(\phi) \\
M_y &= M_0 \, \sin(\alpha) \, \cos(\phi) \\
M_z &= M_0 \, \cos(\alpha)
\end{aligned}
\tag{2.6}
$$

where ϕ is the angle of the precession.

The same coil can also be used for signal detection. By switching off the pulse, the signal induced is proportional to the precessing nuclear magnetization and decays with time, and the original state of equilibrium is reestablished, with the magnetization aligned parallel to the main magnetic field described by two separate relaxation time constants, T_1 and T_2 (Bloch's equations [12]).

$$
\begin{aligned}
\frac{dM_z}{dt} &= \gamma(\vec{M} x \vec{B})_z + \frac{M_0 - M_z}{T_1} \\
\frac{dM_{xy}}{dt} &= \gamma(\vec{M} x \vec{B})_{xy} + \frac{M_{xy}}{T_2}
\end{aligned}
\tag{2.7}
$$

The measured signal is called free induction decay (FID).

Stimulation of the magnetic moments with $\alpha = 90°$ [Eq. (2.5)] leads to a precession in the x/y plane decaying with time. The inhomogeneous dephasing precessing magnetic moments can be refocused with a second pulse ($\alpha = 180°$) at time τ causes a spin echo (SE) at time 2τ (Fig. 2.1).

The mobility of the molecules that contain the nucleus causes relaxation. Each nucleus is surrounded by other magnetic moments that are in constant Brownian motion and therefore produce a continuously changing magnetic perturbation field. The part of the magnetization vector that is

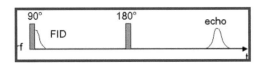

FIGURE 2.1: Spin echo pulse sequence.

parallel to the main magnetic field relaxes with the spin-lattice relaxation time T_1. This is caused by interactions between the observed nucleus and unexcited nuclei in the environment as well as by electric fields in the environment. T_1 is the time required for the longitudinal component of the magnetization vector to be restored to 63% of its original magnitude.

The perpendicular part of the magnetization relaxes with the spin–spin relaxation time T_2. This is caused by interactions between the already excited spinning nuclei. T_2 is the time required for the transverse component of the magnetization vector to drop to 37% of its original magnitude after initial excitation. Nevertheless, T_1 and T_2 are coupled variables. The solution of Bloch's equations is

$$M_{xy}(t) = M_x(t) + iM_y(t) = iM_0 e^{-i\omega t} e^{-t/T_2}$$
$$M_z(t) = M_0(1 - e^{-t/T_1}) \tag{2.8}$$

The measurement of T_2 is only possible in highly homogeneous magnetic fields. In practice, the effective relaxation time T_2^* which is directly related to the main magnetic field deviations ΔB_z is valid

$$\frac{1}{T_2^*} = \frac{1}{T_2} + \frac{1}{2}\gamma \Delta B_z \tag{2.9}$$

Therefore, T_2^* directly measures the local magnetic field inhomogeneities.

2.1.2   Magnetic Resonance Imaging

The required static magnetic field is provided by a superconducting magnet. Typical magnetic field strength for medical applications and human brain research is 1.5 or 3.0 T. The homogeneity is assured by additional field gradients called shim system. The RF sender provides pulses with a frequency $\omega/2\pi$ of about 64 or 128 MHz (proton resonance), respectively. When the gradient system applies gradient pulses in the order of 100 mT/m, the detection system acquires the data (Fig. 2.2).

In order to analyze the signal with respect to its spatial origin, at least one of the two fields required for MR measurement has to vary over space. The simplest way for a two-dimensional (2-D) MR imaging is a standard spin echo sequence (Fig. 2.3).

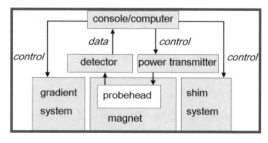

FIGURE 2.2: Block diagram of an MR tomograph.

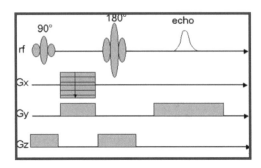

FIGURE 2.3: Typical 2-D slice imaging pulse sequence with spin echo.

Applying a gradient in z-direction during stimulation with an amplitude-modulated RF-pulse causes a slice selection perpendicular to z. The spatial information in x- and y-direction is encoded by a phase-encoding gradient in x-direction and a read-out gradient in y-direction. The pulse sequence has to be repeated for each step of the phase-encoding gradient. The read-out gradient has to be "refocused" before the 180°-pulse. The information is sampled in the reciprocal k-space and is then Fourier transformed into r-space, thereby reproducing the spin-density distribution of the measured object. However, the information consists of complex numbers and thereby the phase of the signal as function of k can be derived, the so-called k-space imaging. This information is used, e.g., in chemical shift imaging [11]. A 3-D image can be recorded by repeating the 2-D sequence for several slices. The respective slice selection is achieved by changing the stimulating RF-frequency.

Two parameters are available for signal manipulation: the sequence repetition time T_R and the echo time T_E. The use of a long T_E allows transverse relaxation of the system to become effective before signal acquisition, whereas, rapid repetition of the pulse sequence prevents longitudinal magnetization from reestablishing. The signal intensity in a picture element (2-D: pixel, 3-D: voxel) is given by

$$M_{xy}(x, y, z) = M_0(x, y, z)(1 - e^{-T_R/T_1(x,y,z)})e^{-T_E/T_2(x,y,z)} \qquad (2.10)$$

The repetition time and the echo time can be adjusted so that the image contrast due to different types of tissue is determined by M_0, T_1 or T_2. Short values of T_R and T_E give T_1 weighted images whereas long T_E and short T_R give spin-density weighted images. Long values of T_R as well as T_E give T_2 weighted images. An example is shown in Fig. 2.4.

Improvement and simplification is done by a gradient echo sequence (Fig. 2.5). Here, short repetition times are combined with small flip angles (α) [13]. Thus, it is possible to record multiple gradient echoes. The possibility of recording images within seconds occurs [echo-planar imaging (EPI)]. As gradient echo sequences are fast, they are well suited for 3-D data acquisition. The sensitivity of gradient echo sequences to magnetic field inhomogeneities can be utilized for image effects in the human body, which responds sensitively to changes in magnetic susceptibility.

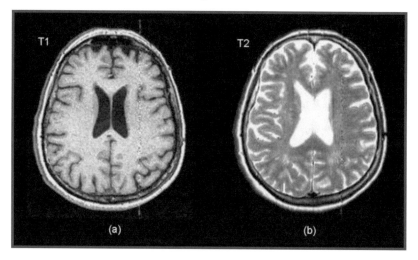

FIGURE 2.4: Example for different image contrasts of appreciatively the same slice of a human brain. (a) T_1-weighted image. (b) T_2-weighted image.

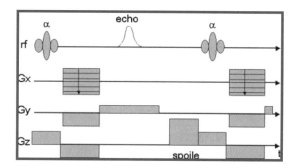

FIGURE 2.5: Gradient echo sequences for fast image acquisition.

For more details about magnetic resonance instrumentation and pulse sequences the reader is referred to [11,9,8,10].

2.1.3   Mathematical Excursion 1: Smoothing, Interpolation and Normalization During Postprocessing

In order to improve the image quality and to unify the image or recording dimensions, several postprocessing procedures can be applied.

2.1.3.1 Smoothing and Interpolation. As different MRI scans can be recorded with different voxel resolutions and often the recording technique provides voxels with nonisotropic size (e.g., the slice thickness is larger than the in-plane voxel resolution or pixel size), the transformation to an isotropic voxel size may be of interest. The standard transformation procedure is a linear nearest neighbor transformation with

$$I_{\text{target}}(i, j, k) = \sum a_v I_v(l, m, n) \tag{2.11}$$

where $I_{\text{target}}(i, j, k)$ is the voxel intensity at the new grid coordinates i,j,k, and l,m,n are the original recorded voxel coordinates in x,y,z direction. $I_v(l, m, n)$ are the voxel intensities of the eight recorded voxels that contribute to the target intensity with factors a_v, the eight weighting factors for the interpolation.

Interpolation and smoothing are classical image postprocessing procedures for which a large variety of methods exist. Mishra et al. [14] developed the idea of an anisotropic image interpolation method. Here, the kernel for interpolation is weighted by a factor depending on the local gradients. Thus, the sharpness of the image is preserved. For smoothing, a Gaussian kernel can be chosen, e.g., a kernel sphere around each voxel with Gaussian shape is used. This kernel is then weighted with the local gradients by

$$I_{\text{target}}(i, j, k) = \frac{\sum I_v(l, m, n)/(r_v + g_{\text{target}})}{\sum 1/(r_v + g_{\text{target}})} \tag{2.12}$$

where r_v is the distance between I_{target} and I_v. g_{target} is the absolute value of the gradient at position (i,j,k). In this way, the local gradients weigh the interpolation kernel with a sharpness dependency.

2.1.3.2 Normalization. The measured intensity of T_1- or T_2-weighted images depends on acquisition parameters and thus does not reflect a measurable value. For comparison of recordings from different subjects or from different scanners, intensity normalization may be helpful. For this task, the average $\bar{\mu}_1$, $\bar{\mu}_2$ and standard deviation σ_1, σ_2 of the signal intensities have to be calculated.

$$\bar{\mu}_i = \frac{1}{N} \sum_{v=1}^{N} I_v \quad \sigma_i^2 = \frac{1}{N} \sum (I_v - \bar{\mu}_i)^2 \tag{2.13}$$

Intensity transformation of data set 1 to the parameters of data set 2 is performed by

$$I_{1,\text{new}}(i, j, k) = \frac{\sigma_2}{\sigma_1}(I_1(i, j, k) + \bar{\mu}_2 - \bar{\mu}_1) \tag{2.14}$$

Note that the normalization can be performed only after transformation to isotropic voxels. It has to be assured that the recorded data contain the same volume – otherwise errors due to different scanning regions or (in case of different subjects) due to different head sizes can occur. It has to be assured that only voxels inside the head are taken into account. Different signal-to-noise (SNR) ratios can also contribute to an error. A detailed consideration of this topic will extend the focus of this booklet and is left to the study of the interested reader.

2.1.4   Functional Magnetic Resonance Imaging

MRI for mapping human brain functionality has rapidly increased during the 1990s. Now, cerebral physiological responses during neural activation can be measured noninvasively [15,16]. fMRI uses the blood oxygenation-level-dependent (BOLD) contrast [17,18,19]. Here, slight physiological alterations, such as neuronal activation resulting in changes of blood flow and blood oxygenation are detected. These signal changes are related to changes in the concentration of deoxyhemoglobin which acts as an intravascular contrast agent for fMRI [20]. The vast majority of fMRI examinations is performed with BOLD-based methods using T_2^*-weighted gradient echo pulse sequences sensitive to local distortions in the magnetic field (susceptibility sensitive techniques) [21,22,23].

Various fMRI methods have good spatial and sufficient temporal resolution, limited by the precision with which the autoregulatory mechanisms of the brain adjust blood flow in space according to the metabolic demands of neuronal activity. Since these methods are completely noninvasive without the need for contrast agents or ionizing radiation, repeated single-subject studies are feasible.

The fMRI experiment consists of a functional template or protocol (e.g., alternating activation and rest for a certain time as the classical "block design"), which induces a functional response in the brain. The aim of an fMRI experiment is to detect this stimulus response, resulting from the BOLD effect, in a defined volume element. During the evaluation of the resulting fMRI image series the physiologically induced signals have to be separated from noise or from artifacts resulting from patient movement or MRI detection techniques. The functional information of a voxel has to be extracted from its functional time course. Therefore, for each functional time point one fMRI volume is recorded. The complete 4-D data set (three dimensions in space, one dimension in time) consists of subsequently recorded 3-D volumes, and thus a functional time course exists for each voxel of a volume. The acquisition of these functional volumes runs over periods lasting up to several minutes. Therefore, it is essential for the following functional analysis to eliminate even very small head movements. Thus, prior to the analysis of a voxel time course, the sequentially acquired volumes have to be corrected for these head motions. The majority of motion correction algorithms found in the appropriate literature is voxel-based matching algorithms [24,25,26].

As a further fMRI specific problem, in most cases, the signal intensity difference between neighbored voxels in a slice is in the range of the expected BOLD effect. Therefore, complex algorithms have to be developed to extract the functional regions. Common techniques for the analysis of a voxel time course are deterministic techniques, i.e., the similarity to the stimulus function is calculated [27,28], statistically analytical (the differences between activation and rest of two samples are calculated by performing Student's t-test, Kolmogorov-Smirnov test, etc.) [27,29], and have an application of clustering techniques, to improve the SNR of the results [27,30,31].

2.1.4.1 Motion Correction.   The functional information is contained in a series of volumes, which are acquired over a certain time period. Thus, prior to the analysis of a voxel time course, the sequentially acquired functional volumes should be corrected for head motion caused by translation and rotation. Motion correction can be performed under the following assumptions:

- The brain is considered as a rigid body that can move slightly with six degrees of freedom.
- There are no relative movements of single parts (the brain is not pulsating).
- The central slice of a 3-D fMRI volume acts as reference slice for the motion correction.
- The first volume of a series of volumes contributing to an fMRI experiment acts as reference coordinate system.

With these prerequisites, the motion correction is performed in the following way:

First, the rotation and shifting matrices $\vec{\vec{R}}$ and $\vec{T}$ are determined for a reference slice of the volume to be corrected

$$\vec{x}^{(2)} = \vec{\vec{R}}\vec{x}^{(1)} + \vec{T} \tag{2.15}$$

where $\vec{x}^{(1)}$ and $\vec{x}^{(2)}$ are the position vectors of voxels before and after transformation, respectively. The Euler angles (ϕ, θ, ψ) define the rotation matrix

$$\vec{\vec{R}} = \begin{pmatrix} \cos\phi\cos\psi - \sin\phi\sin\psi\cos\theta & -\sin\phi\cos\psi - \cos\phi\sin\psi\cos\theta & \sin\psi\sin\theta \\ \cos\phi\sin\psi + \sin\phi\cos\psi\cos\theta & -\sin\phi\sin\psi + \cos\phi\cos\psi\cos\theta & -\cos\psi\sin\theta \\ \sin\phi\sin\theta & \cos\phi\sin\theta & \cos\theta \end{pmatrix}$$

$$\tag{2.16}$$

The fitting can be performed, e.g., either by Levenberg–Marquardt method or by Conjugated Simplex method [32]. Second, the resulting matrices are used for the transformation of the whole 3-D volume using Eq. (2.15).

2.1.4.2 Functional Analysis. The task of functional analysis of block designs is to measure the difference of activation and rest intensities. In Fig. 2.6 a typical fMRI stimulus–response curve in a block design is displayed.

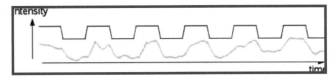

FIGURE 2.6: Stimulus function (dark blue) and response (light blue) in a voxel for a typical fMRI experiment.

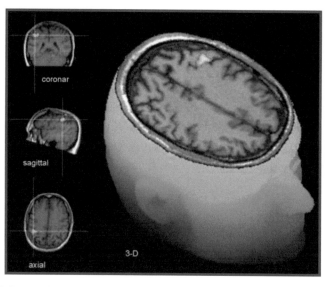

FIGURE 2.7: fMRI detected response from a right hand second finger movement. Activation result is displayed in color. The results are overlaid to a T_1-weighted 3-D image.

In the following excursion, examples for analysis methods are proposed. Using these methods, the resulting activation for a right hand second finger movement is obtained (Fig. 2.7).

Different approaches are event-related designs that associate brain processes with discrete events occurring at any point during the scanning session. The statistical analysis methods are identical for block designs. The deterministic methods can only be applied after a re-sorting of the data.

2.1.5   Mathematical Excursion 2: Are Two Distributions Different?

The decision that two distributions are different can be performed either by deterministic methods or by statistical methods.

2.1.5.1 Deterministic Methods.   The deterministic tools reveal the correlation between the time course and the stimulus function. In fMRI experiments, the measurement protocol requires that the stimulus function be periodic with a time period τ and a frequency $\omega/2\pi$. The total measurement time is then $T = Mt$, and L is the number of time points in the time course, i.e., the number of volumes contributing to the fMRI experiment. $S(t_i)$ is the transformed time course function with average $\langle S(t_i) \rangle = 0$. This is reached by

$$S(t_i) = I(t_i) - \langle I(t_i) \rangle_T \qquad (2.17)$$

where $I(t_i)$ is the functional time course intensity of the appropriate voxel. It is possible to express the transformed time course function in Fourier terms (phase sensitive detection).

$$S(t_i) = \sum S_n \sin(n\omega t_i + \phi_n) \tag{2.18}$$

2.1.5.2 Correlation Analysis with a Sinus Wave Stimulus Function. The correlation of a functional time course to a sinus wave stimulus function is the analogy to the two-phase sensitive detection of a monochromatic signal. This is performed by the multiplication of the functional time course signal with a sinusoidal in-phase reference function

$$R_1(t_i) = \sin(\omega t_i) \tag{2.19}$$

and a sinusoidal out-of-phase reference function (shifted by $\pi/2$):

$$R_2(t_i) = \cos(\omega t_i) \tag{2.20}$$

$$K_{\text{inphase}} = \frac{1}{L} \sum S(t_i) R_1(t_i) \tag{2.21}$$

and

$$K_{\text{outphase}} = \frac{1}{L} \sum S(t_i) R_2(t_i). \tag{2.22}$$

By use of the Fourier expression of $S(t_i)$ (2.18) and the orthogonality of the sinus harmonics

$$\sum_{i=0}^{L} \sin(n\omega t_i) \sin(m\omega t_i) = 0 \quad \forall n \neq m \tag{2.23}$$

two terms are obtained

$$K_{\text{inphase}} = \frac{S_1}{2} \cos(\phi_1) \quad \text{and} \quad K_{\text{outphase}} = \frac{S_1}{2} \sin(\phi_1). \tag{2.24}$$

Now, C_{si} is the result of the correlation analysis component of the amplitude of the spectrum of a sinusoidal time course function.

$$C_{si} = \sqrt{K_{\text{inphase}}^2 + K_{\text{outphase}}^2} = \frac{S_1}{\sqrt{2}} \tag{2.25}$$

or in terms of the functional time course intensity

$$C_{si} = \frac{1}{L} \sqrt{\left(\sum (I(t_i) - \langle I(t_i) \rangle_T) \sin(\omega t_i) \right)^2 + \left(\sum (I(t_i) - \langle I(t_i) \rangle_T) \cos(\omega t_i) \right)^2}. \tag{2.26}$$

C_{si} extracts the monochromatic contribution of the functional time course function. The result is weighted with the functional time course intensity of the appropriate voxel. Due to the analogy of phase sensitive detection, phase shifts between time course and stimulus function are automatically resolved. It is obvious from these calculations that noise contribution can only occur from noise components that fit into the monochromatic contribution, i.e., in the first harmonic.

The calculation of the corresponding significance used in Eq. (2.31) (see below) needs

$$S_{si} = \frac{C_{si}}{\sqrt{\frac{1}{L}\sum\left((I(t_i) - \langle I(t_i)\rangle_T)T(t_i) - \frac{1}{L}\sum(I(t_i) - \langle I(t_i)\rangle_T)T(t_i)\right)^2}} \tag{2.27}$$

where $T(t_i) = 1$ is the activation, $T(t_i) = -1$ is the rest is the stimulus function with average $\langle T(t_i)\rangle_T = 0$ over the time course. Note that the significance reveals the degree of correlation between time course and stimulus and does not take the functional time course intensity into account.

2.1.5.3 Correlation Analysis with a Square Wave Stimulus Function. A square wave stimulus function ($T(t_i) = 1$: activation, $T(t_i) = -1$: rest, $\langle T(t_i)\rangle_T = 0$) in Fourier components is

$$R(t_i) = \frac{4}{\pi}\left(\sin(\omega t_i) + \frac{1}{3}\sin(3\omega t_i) + \frac{1}{5}\sin(5\omega t_i) + \dots\right) \tag{2.28}$$

A correlation analysis reveals

$$C_{sq} = \frac{1}{L}\sum S(t_i)R(t_i) = \sum\frac{S_n}{n}\cos(\phi_n), \quad (n \text{ odd}) \tag{2.29}$$

or in terms of the functional time course intensity

$$C_{sq} = \frac{1}{L}\sum(I(t_i) - \langle I(t_i)\rangle_T)T(t_i). \tag{2.30}$$

C_{sq} is the mathematical analogy to the phase sensitive detection of a signal with a square wave reference function. This function extracts the correlation of the voxel time course function to the stimulus function. The result is weighted with the signal intensity of the time course function. It is obvious that Eq. (2.29) extracts only the sinus components of $S(t_i)$, i.e. due to the fact that the phase of the voxel time course versus the stimulus function is unknown. C_{sq} has to be calculated as the maximum of the function obtained by progressively shifting the stimulus function for each time point of a period. Thus, if the input signal of interest is not a pure sinusoidal signal but has odd harmonics of the frequency $1/\tau$, the output has additional contributions from these odd harmonics.

To test the statistical significance of signals based on this correlation coefficient, s_{sq}, is calculated according to the rules given in [32]

$$S_{sq} = B_{inc}\left(\frac{L-2}{2}, \frac{1}{2}, s_{sql}\right) \tag{2.31}$$

using the abbreviations

$$s_{sql} = \frac{L-2}{L-2+s_{sqn}^2}, \quad s_{sqn} = s_{sq}\sqrt{\frac{L-2}{(1+s_{sq})(1-s_{sq})}}, \quad s_{sq} = \frac{C_{sq}}{\sqrt{\frac{1}{L}\sum(I(t_i) - \langle I(t_i)\rangle_{t_i})T(t_i))^2}}.$$

Here, B_{inc} is the incomplete beta function which returns a number between zero and one, and represents the probability that s_{sq} could be this large or larger just by chance.

Equation (2.31) reveals the significance of the degree of correlation between time course and stimulus and does not take the functional time course intensity into account, whereas Eq. (2.27) calculates the correlation weighted by the intensity of the time course function.

As a complete understanding of all aspects of the BOLD effect could not be reached yet [27], the theoretical functional time course is also not known exactly. Now, C_{sq} extracts all odd harmonics of a measured sinusoidal time course function, including noise that fits into one of the higher harmonics, whereas C_{si} only cares about the ($n=1$)-component of the time course function and of the noise. Therefore, one has to note carefully if C_{sq} or C_{si} is being calculated.

2.1.5.4 Statistical Analysis Methods. The Student's *t*-test S_t calculates the significance of a difference of means of two ensembles (ensemble M – for activated time points (stimulus "on") and ensemble N – rest time points (stimulus "off"), m and n are the corresponding ensemble members).

$$S_t = 1 - B_{inc}\left(\frac{m+n}{2} - 2, \frac{1}{2}, s_{st}\right) \qquad (2.32)$$

with the abbreviations

$$s_{st} = \frac{m+n-2}{m+n-2+t^2}, \quad t = \frac{\overline{M} - \overline{N}}{\sqrt{(\sigma^2(M)/m + \sigma^2(N)/n)}}$$

where $\overline{M}$ and $\overline{N}$ are the average values for each ensemble and σ^2 the variances [32].

Another statistical test is the Kolmogorov–Smirnov test that calculates the significance of the maximum value of the absolute difference between two ensembles, for details see [32]. Another group of statistical tests are the so-called nonparametric tests that replace the measured values by their ranking when they are in a sorted order. For details see [32]. Many other statistical tests can be thought of and it is left to the reader to study them in detail. For detailed information about fMRI experiments, the reader is referred to [33,34, 35,36,37,38,39].

2.1.6 Diffusion Weighted Imaging

The diffusion of water molecules in the presence of a strong magnetic gradient results in an MRI signal loss as a result of the dephasing of spin coherence. The application of a pair of strong gradients

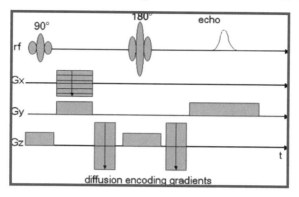

FIGURE 2.8: Typical diffusion encoding pulse sequence.

to detect differences in the diffusivity of water molecules among various biological tissues is known as diffusion weighting [40,23,41]. In diffusion-weighted imaging (DWI), the signal intensity in each voxel S is influenced by the b-value (diffusion sensitization) which is related to the gradient strength G. Diffusion is described by using a single scalar parameter, the diffusion coefficient D. The effect of diffusion on the MRI signal is an attenuation A that depends on D and the b-value

$$A = e^{-Db} \qquad (2.33)$$

Thus, the apparent diffusion coefficient (ADC) for each image voxel can be calculated.

A typical pulse sequence is displayed in Fig. 2.8. 3-D images are recorded for different diffusion encoding gradients (typical b-values are 0, 500 s/mm^2, 1000 s/mm^2). An ADC map of a patient with a lesion in the right hemisphere is displayed in Fig. 2.9. The area of the lesion has reduced diffusion and thus appears dark in the ADC map.

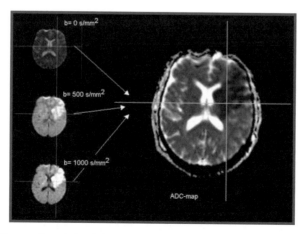

FIGURE 2.9: ADC-map (axial slice) for a patient with a vascular lesion in the right hemisphere.

2.1.7   Diffusion Tensor Imaging (DTI)

The advanced extension of DWI that addresses the visualization of white matter networks is DTI. In DTI at least six gradient directions are recorded to scan a diffusion matrix for each voxel. In this way, not only the diffusion amplitude (as in DWI) is mapped, but also information about the orientational dependence can be obtained.

DTI can be used to characterize the orientational dependence of ^{1}H diffusion in human brain white matter [42,43]. The measured diffusion is apparently maximal along the fiber direction and restricted in the perpendicular direction. ^{1}H diffusion can be mapped spatially, and the observation that this diffusion is reduced in acutely ischemic brain tissue as well as in degenerated brain tissue, is responsible for the increasingly widespread use of these techniques in clinical neuroimaging [44,45, Nimski, 2005c, 46,47].

However, in the presence of anisotropy in white matter, diffusion can no longer be characterized by a single scalar coefficient [40,48], but requires a tensor $\vec{\vec{D}}$, which fully describes molecular mobility along each direction and correlation between these directions [49,50]. Diffusion anisotropy is mainly caused by the orientation of fiber tracts in white matter and is influenced by its micro- and macrostructural features. Of the microstructural features, intraaxonal organization appears to be of major influence on diffusion anisotropy. Other features include the density of fiber and cell packing, degree of myelination, and individual fiber diameter. On a macroscopic scale, the variability in the orientation of all white matter tracts in an imaging voxel influences the degree of anisotropy assigned to the respective voxel [51,52].

The elements of the symmetric tensor $\vec{\vec{D}}$ can be measured by diffusion gradients along at least six noncollinear directions so that b [Eq. (2.33)] has become a tensor, resulting in signal attenuation

$$\ln(A) = -(b_{xx}D_{xx} + 2b_{xy}D_{xy} + 2b_{xz}D_{xz} + 2b_{yz}D_{yz} + b_{yy}D_{yy} + b_{zz}D_{zz}) \qquad (2.34)$$

This requires taking into account possible interactions between imaging and diffusion gradients and even between imaging gradients that are applied in orthogonal directions (cross terms) [53,54].

The second-rank diffusion tensor $\vec{\vec{D}}$ can be diagonalized, leaving only three nonzero elements along the main diagonal of the tensor, the eigenvalues ($\lambda_1, \lambda_2, \lambda_3$). The eigenvalues reflect the shape or configuration of the spheroid. The mathematical relationship between the principal coordinates of the spheroid and the laboratory frame is described by the eigenvectors ($\vec{v}_1, \vec{v}_2, \vec{v}_3$) (Fig. 2.10).

2.1.7.1 Eddy Current Correction. DTI data acquired with echo-planar imaging technique are highly sensitive to eddy current-induced geometric distortions that vary with the magnitude and direction of the diffusion sensitizing gradients. This distortion has a greater effect on the images than the subject head movement and has to be corrected prior to a motion correction as described earlier. For the correction of this distortion, several methods were proposed. The technique of Shen et al.

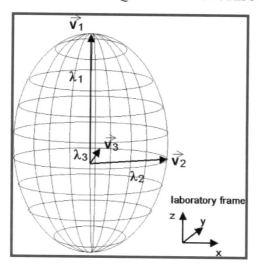

FIGURE 2.10: Diffusion spheroid (schematically) with three eigenvectors ($\vec{v}_1$, $\vec{v}_2$, $\vec{v}_3$) and three eigenvalues (λ_1, λ_2, λ_3). The *e*igenvectors usually do not coincide with the Cartesian frame (x,y,z).

[55] relies on collecting pairs of images with diffusion sensitizing gradients reversed – these paired images are distorted with eddy currents in opposite directions. A columnwise correction in the image domain along the phase encoding direction (anterior–posterior) was applied. This was performed by searching for the maximum value of the cross-correlation between two corresponding columns (of two paired volumes) while one is shifted and scaled (fitting routine – Simplex method [32]). Each column was then corrected by applying opposite shifts and scales equal to half of the correction.

As the recording technique provides voxels with nonisotropic size, the DTI data sets are transformed as a first analysis step into an isotropic grid with voxel size 1.0 mm × 1.0 mm × 1.0 mm. The transformation was performed according to Eq. (2.11).

2.1.7.2 Fractional Anisotropy (FA) Mapping. The diffusion anisotropy can be quantified by several indices [56], each of which distinguishes between a prolonged spheroid (which would be expected in oriented fiber bundles of white matter) and a sphere (which represents isotropic diffusion). Out of these, the fractional anisotropy (FA) is calculated

$$F_f = \sqrt{\frac{(\lambda_1 - \bar{\lambda})^2 + (\lambda_2 - \bar{\lambda})^2 + (\lambda_3 - \bar{\lambda})^2}{\lambda_1^2 + \lambda_2^2 + \lambda_3^2}} \qquad (2.35)$$

where $\bar{\lambda}$ is the arithmetic average of the three eigenvalues. After the eddy current correction and an optional rigid brain motion correction, the FA maps can be calculated (Fig. 2.11).

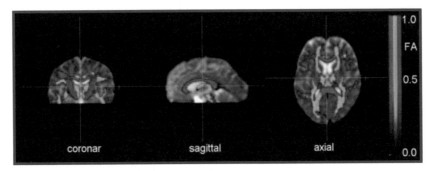

FIGURE 2.11: FA maps of a healthy subject, aged 36 years. The FA maps were overlaid on the b_0 image (for the display a threshold on the FA-values of 0.2 was applied). The color coding was as follows: red for major eigenvector mainly in left-right direction, blue for major eigenvector mainly in inferior-superior direction, and green for major eigenvector mainly in posterior-anterior direction.

2.1.7.3 Fiber Tracking (FT). Insights into the link between functional brain regions and anatomical fiber connections are essential for an integrated understanding of the organization of the human brain and central nervous system. For MRI-based fiber tracking (FT) anisotropic diffusion is characterized to determine the preferred diffusion direction. By calculation of the diffusion spheroid, the eigenvector corresponding to the major eigenvalue is seen as the direction of fastest diffusion and indicates the fiber direction in white matter regions (see scheme in Fig. 2.12).

Based on this directional information, different methods and algorithms have been proposed to estimate white matter connectivity [57,58,59,60,61]:

- *Streamline tracking technique (STT)*: It models the propagation in the major Eigenvector field of the brain.
- *Tensor deflection (TEND)*: It uses the entire diffusion tensor $\vec{\vec{D}}$ to deflect the incoming vector [57] that results in the outgoing vector (2.36) $\vec{v}_{out} = \vec{\vec{D}}\vec{v}_{in}$

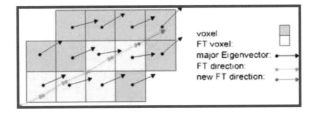

FIGURE 2.12: Scheme for FT. Following the average of surrounding major eigenvectors (black arrows) directions step by step, a voxel is classified as belonging to FT (light yellow) or not (light blue), depending on whether the seed point of the FT direction (green arrow or red arrow, respectively) is at the voxel site.

- *Tensorline algorithm*: It uses STT for highly prolate diffusion tensors and TEND for the diffusion tensors which have a more oblate or spherical tensor shape.

$$\vec{v}_{\text{out}} = f\vec{v}_1 + (1-f)((1-g)\vec{v}_{\text{in}} + \vec{\vec{D}}\vec{v}_{\text{in}}) \qquad (2.37)$$

where f and g are user-defined weighting factors that vary between 0 and 1. The algorithm has three terms: (a) an STT term (the direction of the major Eigenvector is weighted by f), (b) a TEND term weighted by $(1-f)g$, and (c) an undeviated term weighted by $(1-f)(1-g)$. All vectors have to be normalized to unity before they can be used in Eq. (2.37).

Probabilistic methods – the global connectivity is estimated by the calculation of the uncertainty in the direction of the eigenvectors at each voxel. The best estimate of probability density function at all voxels of the track is selected as the fiber track [62].

In our observations, a set of parameters had to be adjusted for a reasonable FT, and the validation of the result was the plausibility judged by an experienced analyst. Parameters to be adjusted were as follows:

- *Threshold for the scalar product*: The threshold for the scalar product of the first eigenvectors (angle between directions) of two consecutive FT positions: not necessarily voxels, since FT positions result from float numbers depending on the step width (i.e., the distance between two FT positions) and the first eigenvector directions is usually about 0.9.
- *Choice between STT, TEND, or the combination of both*: The choice between STT, TEND, or the combination of both – in the latter, the parameters f and g had to be adjusted.
- *The step width*: The step width (distance between two consecutive FT positions) for a reasonable FT, variations of the step width between 0.5 mm and 1.0 mm are possible.

FT results are displayed in Fig. 2.13 (TEND, step width 0.5 mm, threshold 0.9).

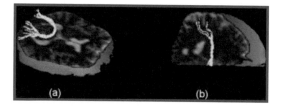

FIGURE 2.13: Fiber tracking results for a healthy subject – the images are glass-brain plots of the FTs overlaid on 3-D views of $(b = 0)$-data: (a) starting point in the dorsal corpus callosum, (b) starting point in the left internal capsule.

2.1.8   Magnetic Resonance Spectroscopy (MRS)

MRS takes advantage of the fact that protons reside in different chemical environments, depending on the molecular structure. The distribution of these protons can be displayed as a spectrum with the main peak composed of water and other peaks depending on the molecular structure. Nevertheless, the multimodal integration MRS is mathematically similar to the integration of fMRI or DTI and thus the focus of this synopsis is not set to MRS.

2.2   MAGNETOENCEPHALOGRAPHY

2.2.1   Principles of Magnetoencephalography

Biomagnetic signals from the human brain are extremely weak in comparison to the earth's magnetic field (5×10^{-5} T) or magnetic urban noise: they are in the order of femtotesla (10^{-15} T) with frequency between 1 Hz and 1 kHz. Thus, highly sensitive sensors are needed and the influence of disturbing magnetic fields has to be reduced.

This sensor is the superconducting quantum interference device (SQUID). A SQUID sensor uses the phenomenon of superconductivity, i.e., a superconductor loses its electric dc resistance below a certain temperature and a magnetic field is expelled from the interior of a superconductor (Meissner–Ochsenfeld effect). The magnetic flux Φ_{sc} caused by the external magnetic field B enclosed by a superconducting loop with the area A is kept constant:

$$\Phi_{sc} = BA = const. \tag{2.38}$$

The magnetic flux is quantified

$$\Phi_{sc} = n\Phi_0 = n\hbar/2e \tag{2.39}$$

where n is an integer, $\hbar$ is the Planck's constant, e is the elementary charge, and Φ_0 is the magnetic flux quantum.

To keep the magnetic flux constant, a screening current I_{sc} is kept flowing through the superconducting circuit when the external magnetic field has changed. Two main classes of SQUIDs depending on the operating principle exist: dc SQUIDs consists of two weak links in the circuit whereas RF SQUIDs have only one. For dc SQUIDs, if a dc bias current is applied to the SQUID loop, then a voltage U can be measured. This voltage is sensitive to the external magnetic field and thus a SQUID represents a flux to voltage converter.

The SQUID operates below 4.2 K, i.e., is usually placed in liquid helium. The typical area covered by a SQUID chip is 4×4 mm^2 and the field sensitivity of a biomagnetic measurement system can be expressed by flux density noise in $f\,T/\sqrt{Hz}$. The best values achieved were $2.5 f\,T/\sqrt{Hz}$. To measure the spatial magnetic field distribution/dependence, an array of SQUIDs is placed in a Dewar under superconducting conditions. To obtain a better SNR by reducing the ambient disturbances, the recording system is placed in a magnetically shielded room. The shielding is based

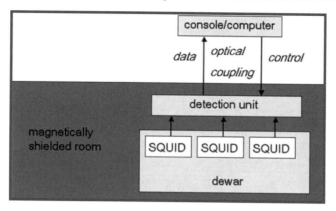

FIGURE 2.14: Schematical MEG recording device. An array of SQUIDs is placed in a Dewar. The whole detector system is operating inside a magnetically shielded room. The coupling to the console is optically controlled.

on diversion of the flux by means of layers of metal with low magnetic permeability. Special sensor configurations are additionally used to reduce signal contributions from homogeneous magnetic disturbances. Figure 2.14 shows a schematic overview of an MEG device.

The MEG signals are derived from the net effect of ionic currents flowing in the dendrites of neurons during synaptic transmission. In accordance with Maxwell's equations, any electrical current will produce an orthogonally oriented magnetic field, which can be measured by MEG. The net currents can be thought of as current dipoles, which are currents defined to have an associated position, orientation, and magnitude, but no spatial extent. According to the right-hand rule, a current dipole gives rise to a magnetic field that flows around the axis of its vector component.

In order to generate a signal that is detectable, approximately 50,000 active neurons are needed. Since current dipoles must have similar orientations to generate magnetic fields that reinforce each other, it is often the layer of pyramidal cells in the cortex, which are generally perpendicular to its surface, that give rise to measurable magnetic fields. Furthermore, it is often bundles of these neurons located in the sulci of the cortex with orientations parallel to the surface of the head that project measurable portions of their magnetic fields outside of the head. For more details about biomagnetic instrumentation refer to [63,27].

Examples of MEG recordings are shown in Fig. 2.15. The task performed was a right hand second finger movement. Due to the low SNR, single-event display is not useful, so (see Chapter 4) the events were detected and averaged according to trigger myogram. Other examples for MEG are given in [64,65,66].

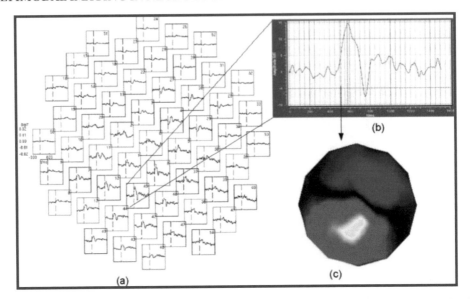

FIGURE 2.15: Averaged MEG event. (a) Sensor distribution plot – the averaged event is displayed at the respective sensor position. (b) Single channel recording of a brain event. (c) Magnetic field distribution for a certain time point marked as line in (b) (yellow-red: positive magnetic field, blue: negative magnetic field).

2.2.2   Source Reconstruction

In order to determine the location of the activity within the brain, advanced signal processing techniques that use the magnetic fields measured outside the head to estimate the location of the source of that activity are used. This is referred to as the inverse problem.

To represent the simplest kind of source, a current dipole is often used to explain the relationship between neuronal activity and the measured electric and magnetic fields. The current dipole $\vec{q}$ is a point source, a mathematical model to represent a short element of current. A current density within a volume conductor Ω induces a magnetic field outside. When the conductivity σ within Ω and the electric current are known, Maxwell's equations and the continuity equation

$$\nabla \cdot \vec{J} = -\frac{\partial \rho}{\partial t} \qquad (2.40)$$

where $\vec{J}$ is the current density and ρ is the charge density can be used to calculate the electric field $\vec{E}$ and the magnetic field $\vec{B}$. The current dipole is related to the primary current density $\vec{J}_p$ in a differential volume element dv.

$$\vec{q} = \int \vec{J}_p dv \quad \text{or} \quad \vec{J}_p(r) = \vec{q}\delta \qquad (2.41)$$

The total current density can be divided into two components

$$\vec{J} = \vec{J}_p + \vec{J}_\Omega \qquad (2.42)$$

with $\vec{J}_\Omega = \sigma \vec{E} = -\sigma \nabla V$ is the volume current which results from the effect of electric field on charge carriers in the conducting medium. Solving Eqs. (2.40–2.42) together with Maxwell's relations, a relation between the measured magnetic field and the current density is obtained

$$\nabla^2 \vec{B} = -\mu_0 \nabla \times \vec{J} \qquad (2.43)$$

with $\vec{B} = 0$ at ∞.

Equation (2.43) is of Poisson's type. Analytical solutions can be found only for simple geometries and conductivity distributions, i.e., volume conductors like half space, sphere, or ellipsoid. The reader interested in the theory of solution of the forward and inverse problems is referred to [63,67].

In the example presented here, the MEG source reconstruction process is performed using single dipole representations of the underlying source by the Boundary Element Method: for a piecewise homogeneous volume conductor, consisting of N homogeneous and isotropic compartments, the magnetic field generated by an active source can be calculated by the Geselowitz's formula [68,63]. The global field is given by the contribution of the infinite medium magnetic field and by the secondary sources represented by the electric potential $V(\vec{r})$ at the conductivity interfaces s_0 and s. An integral equation describes the potential on each surface between compartments. The homogeneous single-compartment conductor model, usually sufficient for MEG [67] is used in the following. A tessellation of the surface into an appropriate number of triangles is needed, and the potential values at each of these triangles are approximated using a collocation method with linear basis functions [69]. This means that the integral equation for the potential must be satisfied only at the M node points represented by the vertices of the triangles where the potential V is calculated from

$$\left(I - \frac{1}{2\pi} \Omega \right) V = \frac{2\sigma_0}{\sigma} V_0 \qquad (2.44)$$

where V_0 is the vector of the infinite medium potential at the M nodes and W the matrix of the solid angles.

Once the numerical expression of the secondary source is obtained, the global magnetic field is analytically calculated as a linear combination of vector functions $a_i(r)$ [70,71]

$$B(r) = B_0(r) + \sum_{i=1}^{M} V_i a_i(r) \qquad (2.45)$$

A dipole localization at a given time instant in the averaged data can be performed using either a linear iterative algorithm [72] or a simplex conjugate algorithm for double check. The cost function is based on the match of the measured map with the predicted map for the trial solution according to a least squares approach or a correlation-based criterion

$$J_{LS} = \left\| \vec{B} - \overrightarrow{LQ^T} \right\|^2 \tag{2.46}$$

where $\vec{B}$ is the measured signal map and $\vec{B}_{cal} = \overrightarrow{LQ^T}$ is the predicted map calculated from the estimated dipole parameters.

A linear iterative algorithm for inverse calculation [72] consists of a least squares parameters estimation combined with a first-order approximation of the field variation corresponding to a given parameter increment. This method linearizes locally, the problem to calculate increments for position Dr and momentum Dd thus transforming the standard nonlinear least square parameter estimation into a linear iteration scheme (computationally similar to a Levenberg–Marquardt approach).

The simplex conjugate algorithm is a standard nonlinear least squares parameter estimation routine, an algorithm with randomly distributed starting points for each iteration is used to estimate the position of the dipole.

Generally, the localization procedure inside the whole localization algorithm is performed twice: first, an analytically represented conductor is used as volume conductor model (spherical model for total head MEG) and in this way an approximation of the final dipole position is achieved. The second run of the localization algorithm is performed using the actual volume conductor (realistic shaped volume conductor or its ellipsoid approximation), and a refinement of the previous solution is found. The use of the ellipsoid approximation can be used as an alternative to the realistic shaped volume conductor; nevertheless, for the dipole localizations presented here no significant differences could be observed.

Both of the algorithms (nonlinear minimization or its linearized version) provide a solution for the dipole location. The three parameters given by the position of the dipole have, in fact, a nonlinear dependency on the field due to the lead field matrix relating the field to the dipole position [68]. The strength $\vec{Q}$ of the localized source is, instead, linearly related to the field and can thus be obtained through direct pseudoinverse solution, given the location of the dipole. For both of the inverse algorithms, the linear system, derived from the maximum likelihood approach, is solved to get the momentum components $\vec{Q}$ once the position has been determined:

$$\left(\vec{\bar{L}}^T \vec{\bar{L}} \right) \vec{Q} = \vec{\bar{L}}^T \vec{B} \tag{2.47}$$

where $\vec{\bar{L}}$ is the lead field matrix and $\vec{B}$ is the matrix $3 \times N_s$ of the components of the field at every measurement site (N_i is the number of channels of the MEG system).

2.2.3   Modeling

To obtain the structural background for MEG source reconstruction, 3-D models from MRI data are generated and used as a morphological basis for the MSI reconstruction. Tissue segmentation is performed using the voxel intensity above or below respective thresholds.

Tissue segmentation is a topic already covered extensively in scientific literature. Thus, in the framework of this study, the strategy to obtain the inner skull is explained in short. The outer skin can be obtained from MP-RAGE data sets by setting up ray traces from the image borders to the center of the head, and subsequently (starting from the image borders and going to the center of the head) defining the outer skin as the first point where the image voxel intensity is higher than the noise threshold.

Alternatively, edge detection algorithms can be used. The simplest one is to use the derivative instead of the original data.

$$D(i, j, k) = \frac{1}{2}|I_{(i+1,j,k)} - I_{(i-1,j,k)}| + \frac{1}{2}|I_{(i,j+1,k)} - I_{(i,j-1,k)}| + \frac{1}{2}|(i,j,k+1) - I_{(i,j,k-1)}| \qquad (2.48)$$

Instead of using the derivative distance of one, two or three voxels or a combination of them can be used.

Traveling further along a ray trace, the inner skull then can be derived as the minimum of the derivative of the voxel intensity values. This definition properly works for regions that are relevant for MEG (e.g., parietal cortex of the hemisphere). For example, eye holes, mouth, etc. are not relevant for the brain source reconstruction, and there is no need for a high-resolution segmentation of those head regions. Finally, the brain tissue can be segmented by a region-growing algorithm working with simple thresholds that need to be adjusted.

A threshold-dependent region grow algorithm follows the following iteration process:

1. Mark a starting point.

2. Test if the intensities of the neighbored voxels to a marked voxel are in the range adjusted by the thresholds.

3. Mark the voxels.

4. Go to 1 till no further voxel is marked.

The triangulation of the resulting tissue components is performed by expanding an already triangulated sphere positioned in the center of the respective tissue until the tissue boundaries are reached for each triangle edge separately, thus rendering a realistic volume conductor model. Additionally, for use in realistic models which have a complex boundary and in the case of a very superficially located source – in order to avoid error due to tessellation – an ellipsoid could be used which is fitted with its center, length, and orientation to the axis to the realistic model. The realistic volume conductor used for source reconstruction is the triangulated inner skull.

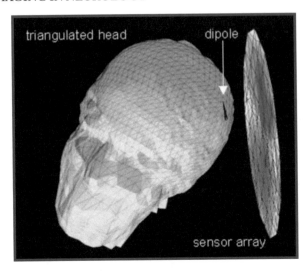

FIGURE 2.16: 3-D visualization of a triangulated head surface (about 1700 triangles) with dipole localization result displayed as a pin. On the left side, a sensor array with magnetic field distribution for a certain time point is displayed schematically.

2.2.4   Dipole Localization

Using the model described in Section 2.3 and with the fitting procedure described in Section 2.2 an example of a single dipole localization for the magnetic field distribution at a defined time point is calculated (Fig. 2.16).

2.3   OTHER TECHNIQUES

To add quantitative information of regionally disturbed brain function or regional tumor metabolism into the common framework of image analysis and display, the integration of PET and SPECT into multimodal functional neuroimaging can be used. Generally, PET as well as SPECT generates images of the regional and temporal distribution of a radio-labeled pharmaceutical after injection. For SPECT, pharmaceuticals are usually labeled with the photon emitters ^{99m}Tc or ^{123}I, which due to their radiochemical properties and their half-life of 6 hours and 13 hours, respectively, allow either a daily on-site preparation of the radiopharmaceuticals or a delivery from a regional provider. The positron emitters needed for PET, however, have half-lives ranging from 2 hours (^{18}F) to 2 minutes (^{15}O) so that for all preparations of radiopharmaceuticals apart from ^{18}F labeled compounds, a local cyclotron facility is mandatory.

The image resolution, depending on the technique and instrumentation, is ranging from a spatial resolution of 10–15 mm for SPECT to 4 mm for PET [7].

PET and SPECT as techniques for scientific investigation in humans are limited by the need for clearance by ethics committee to inject radioactive material into participants. The minimization of radiation dose to the subject is an attractive feature of the use of short-lived radionuclides.

In short, the raw data collected by a PET scanner are a list of "coincidence events" representing near-simultaneous detection of annihilation photons by a pair of detectors. Each coincidence event represents a line in space connecting the two detectors along which the positron emission occurred.

Coincidence events can be grouped into projections images called sinograms. The sinograms are sorted by the angle of each view and tilt, the latter in 3-D case images. A normal PET data set has millions of counts for the whole acquisition.

A detailed overview and technical details for PET and SPECT is beyond the scope of this synopsis, which puts special focus on different neuroimaging techniques.

Integration into multimodal imaging is performed in analogy to the other techniques. Using PET specific markers and detecting them by a marker detection system (Chapter 3) allows for easy integration of the results [7].

Another brain mapping technique performing the neurophysiologic measurement of the electrical activity of the brain is called EEG. EEGs are frequently used in experimentation because the process is noninvasive. EEG is performed by recordings from electrodes placed on the scalp, or subdurally, or in the cerebral cortex. The resulting traces represent an electrical signal (postsynaptic potentials) from a large number of neurons. Really, not electrical currents are measured, but voltage differences between different parts of the brain.

EEG is capable of detecting changes in electrical activity in the brain with a very high temporal resolution (of the order of milliseconds). In conventional scalp EEG, the recording is obtained by placing electrodes on the scalp after application of a conductive gel to reduce impedance.

To perform dipole or current source reconstruction as in MEG, a map of electrode signals has to be recorded. This cannot be performed by placing single electrodes, but the subject has to wear a plastic cap where the electrodes are inserted in small holes. Electrode placement is determined by measuring and marking the scalp using a system of placement that is reliable and reproducible (see Chapter 3).

The limitation in contrast to MEG is that MEG has fixed sensor positions but EEG sensor positions vary from each recording. Thus, for current source reconstruction, each electrode position has to be detected before the measurement which can be a long-lasting process or might even result in inaccuracy of the results.

From an economical point of view EEG should be preferred to MEG but the technical point of view tells us superiority of MEG versus EEG. In the following, EEG will not be presented in detail, as well as another brain mapping technique, i.e., transcranial magnetic stimulation.

CHAPTER 3

Coordinate Transformation

3.1 TRANSFORMATION IN INTERMODAL MULTIMODALITY

Multimodal integration requires the 3-D registration of all involved modalities in one common coordinate frame. As it is usually not possible to detect modality specific markers in the environment of any other modality, problems arise with multimodal mapping techniques. Moreover, specific measurement modalities often need different marker locations.

MRI results have a high spatial resolution and if no high temporal resolution is required, they can stand alone. On the other hand, MEG (and EEG) mapping ask for the combination of the resulting data with the anatomical data provided by MRI. Simple marker substitution for the appropriate modality at identical positions introduces an additional error source to the overall system. Simultaneous detection of different types of markers in an optical manner and registration in a common 3-D coordinate frame overcomes the outlined problem.

If the multimodality is based on two imaging modalities a different approach is possible. Instead of transforming from one coordinate frame to another, it is possible to transform all image information from each modality to a stereotactic standard space.

3.1.1 Tracking Systems

Different tracking systems based on different physical principles are available. Usually, tracking systems allow for detection of position and orientation. For multimodal imaging, the orientation detection is not necessary, the position of the modality specific markers is the only parameter needed. Tracking systems work with a sender and a receiver unit: the detector unit has to be placed at the appropriate position, and the sender detects the receiver position.

The electromagnetic tracking system contains three coils with perpendicular axes producing an electromagnetic ac field, which is inducing currents in the receiver. In this way, the actual position of the receiver can be calculated. The problem is of the metal in the surrounding of the system that might disturb the measurement because the electromagnetic fields cause currents in these metals. To overcome this problem, dc currents can be used. These and other disturbances such as other ferromagnetic materials and the earth's magnetic field have been overcome by shielding and earth field compensation (Polhemus, Colchester, Vermont, USA [73]).

An acoustical tracking system is based on ultrasound sender and receiver, which is different in the physical principle but has the same mathematical basis for calculating the positions.

The disadvantage of these tracking systems is that the recording time increases with an increasing number of markers and errors due to patient movements that are difficult to avoid. Improvement is possible by optical systems where the recording time is less than 1 s independent of the number of markers.

3.1.2   Optical Systems

Optical systems have no need for a sender unit, since the optical image recorded by a camera contains all information needed. To evaluate the position of an object, it has to be recorded from at least two cameras where the position and the orientation of the camera axes have to be known exactly. A good example is the OptiCoS [74]. In the following, the detection of marker positions and the appropriate coordinate transformations are explained on the basis of this system.

The system utilizes color video cameras, homogeneously distributed over a 1000 mm diameter circle line at equal distance (Fig. 3.1). The overlapping of the recording segments allows for a 3-D reconstruction by transforming the 2-D coordinates of the appropriate camera image into 3-D coordinates. This transformation is possible as the camera locations are defined and the optical axis of each camera is tilt against the circle plane by an angle of 45° thus focusing a common point 500 mm distant.

The cameras are working with a resolution of 768×576 pixels providing approximately 0.5 mm resolution within the investigation field. Detecting a marker means determining the position of its optical center within the corresponding camera image.

FIGURE 3.1: OptiCoS setup with superimposed coordinate frame and lines representing camera locations.

3.1.3   Mathematical Excursion 3: Calculation of 3-D Coordinates from 2-D Images

Each marker position detected by a camera can be expressed by a line connecting the center of the associated camera lens and the center of the marker. Therefore, the 3-D positions of the markers can be determined by establishing these line equations and calculating their intersection points.

In local 2-D coordinate frame of a camera, a marker vector $\vec{a}$ is defined by the pixel column address a_x and pixel row address a_y and the focal distance a_z between lens and image plane. Each camera coordinate frame can be represented globally by its origin vector $\vec{d}_i$ (center of camera lens) and its Euler angles $(\phi_i, \theta_i, \psi_i)$. The latter describes the revolutions necessary for the local coordinate frames to be transformed to align with the global coordinate frame. The geometric conditions resulting from the chosen mechanical setup facilitate these transformations (Table 3.1). Local marker vectors $\vec{a}$ can be transformed to global coordinate vectors $\vec{a}'$ using the matrix $\vec{\vec{R}}_i$ (i: camera, k: marker).

$$\vec{a}'_{ik} = \vec{\vec{R}}_i \cdot \vec{a}_{ik} \tag{3.1}$$

$$\vec{\vec{R}}_i = \begin{pmatrix} \cos\phi_i\cos\psi_i - \sin\phi_i\sin\psi_i\cos\theta_i & -\sin\phi_i\cos\psi_i - \cos\phi_i\sin\psi_i\cos\theta_i & \sin\psi_i\sin\theta_i \\ \cos\phi_i\sin\psi_i + \sin\phi_i\cos\psi_i\cos\theta_i & -\sin\phi_i\sin\psi_i + \cos\phi_i\cos\psi_i\cos\theta_i & -\cos\psi_i\sin\theta_i \\ \sin\phi_i\sin\theta_i & \cos\phi_i\sin\theta_i & \cos\theta_i \end{pmatrix} \tag{3.2}$$

After this transformation, the lines between the cameras and the marker can be described generally

$$p = \vec{d}'_p + \kappa \cdot \vec{a}'_p \quad \text{and} \quad q = \vec{d}'_q + \nu \cdot \vec{a}'_q \tag{3.3}$$

The markers 3-D coordinates are provided by the intersection point $(\vec{p} = \vec{q})$. Due to the limited accuracy of the optical components two lines p and q associated with the same marker but with different cameras will not likely intersect at a common point but rather approach very close.

TABLE 3.1: Euler angles for transformation

	ψ_i Global z axis	θ_i Modified global x axis	ϕ_i Modified global z axis
Camera 0	$\pi/6$	$-\pi/4$	0
Camera 1	$\pi/2$	$-\pi/4$	0
Camera 2	$-\pi/6$	$\pi/4$	π
Camera 3	$\pi/6$	$\pi/4$	π
Camera 4	$\pi/2$	$\pi/4$	π
Camera 5	$-\pi/6$	$-\pi/4$	0

Mathematically, the intersection point is found at half the minimum distance between the lines. Let $\vec{a}_p$ and $\vec{a}_q$ be normalized, let

$$\vec{\Delta} = \vec{d}_p - \vec{d}_q \tag{3.4}$$

and

$$\cos(\alpha) = \vec{a}_q \cdot \vec{a}_p \tag{3.5}$$

then κ and ν can be expressed in closed form as

$$\begin{pmatrix} \kappa \\ \nu \end{pmatrix} = \frac{1}{\sin^2(\alpha)} \begin{pmatrix} 1 & \cos(\alpha) \\ \cos(\alpha) & 1 \end{pmatrix} \begin{pmatrix} -\vec{\Delta} \cdot \vec{a}_p \\ \vec{\Delta} \cdot \vec{a}_q \end{pmatrix} \tag{3.6}$$

The midpoint can be expressed along the closest connection between the two lines as

$$\vec{d}_p + \kappa \vec{a}_p - \frac{1}{2}\vec{s} \tag{3.7}$$

where

$$\vec{s} = \frac{\vec{\Delta} \cdot (\vec{a}_p \times \vec{a}_q)}{|\vec{a}_p \times \vec{a}_q|^2} \vec{a}_p \times \vec{a}_q \tag{3.8}$$

is the vector which connects the two lines with minimal distance. Experience has shown that an intersection point can be assumed if $|s| < 1$ mm.

3.1.4   Mathematical Excursion 4: Coordinate Transformation for Intermodal Multimodality

The following coordinate transformation is valid for all markers independent on the marker detection system (Polhemus, Opticos, etc.). Thus data of different modalities are transformed into one unique coordinate frame.

In this, the coordinates of the first modality markers are known as relative positions of the markers of the second modality. Subsequently, following transformations of three coordinate frames are to be performed

- The coordinate frame of the first modality measurement device (x, y, z).

- The coordinate frame of the marker detection system (x', y', z').

- The coordinate frame of the second modality measurement device (x'', y'', z'').

First, from the marker detection of the first modality measurement, the positions $(\vec{p})$ are known and rotation and shifting matrices $\vec{R}_1$ and $\vec{T}_1$ are determined for the transformation from marker detection system coordinate frame ($\vec{p}'$) to first modality coordinate frame (suggested method – Conjugated Simplex [32])

$$\vec{p} = \vec{\bar{R}}_1\vec{p}' + \vec{T}_1 \qquad (3.9)$$

Second, from marker detection in the second modality ($\vec{m}''$) the rotation and shifting matrices $\vec{\bar{R}}_2$ and $\vec{T}_2$ are determined for the transformation from second modality coordinate frame to marker detection system coordinate frame ($\vec{m}'$) (suggested method: Conjugated Simplex [32])

$$\vec{m}' = \vec{\bar{R}}_2\vec{m}'' + \vec{T}_2 \qquad (3.10)$$

Then, any coordinate in the second modality coordinate frame ($\vec{a}''$) (e.g., position of functional active voxel) can be transformed to the coordinate frame of the first modality by

$$\vec{a} = \vec{\bar{R}}_1(\vec{\bar{R}}_2\vec{a}'' + \vec{T}_2) + \vec{T}_1 \qquad (3.11)$$

3.2   TRANSFORMATION TO A STEREOTACTIC STANDARD SPACE – INTRAMODAL MULTIMODALITY

The fundamental advantage of using spatial normalization is that results obtained from different subjects can be averaged to finally perform a comparison of groups of patients with certain disorders and healthy subjects. Talairach and Tournoux [75] suggested a transformation algorithm to a standard atlas involving the identification of various brain landmarks and piecemeal scaling of brain quadrants. An alternative approach was to use automated brain registration algorithms [76, 26]. Here, a spatial normalization to the Montreal Neurological Institute (MNI) stereotactic standard space via a scanner- and sequence-specific template data set is presented. The MNI defined a new standard brain by using a large series of MRI scans of normal controls. This resulted in the MNI atlas [77].

An example for standard space normalization of MRI data is explained in the following using a template-specific normalization. The template used for the normalization was created in two iterative steps.

3.2.1   First Iteration Step

From each modality (T_1 weighted, T_2 weighted or functional) one volume is arranged with eight operator-defined landmarks [anterior commissure (AC), posterior commissure (PC), vertical PC plane (VCP) superior, VCP inferior, prolonged AC-PC front, prolonged AC-PC back, prolonged PC-left hemisphere border, prolonged PC-right hemisphere border, see Fig. 3.2]. According to these landmarks, an affine transformation (transformation, rotation, dilation) into the MNI space is performed. Template 1 ($A_{\text{templ},1}$) is then created by arithmetic averaging of the voxel intensities of the transformed volumes of single subjects. Thus, a template is created for each modality.

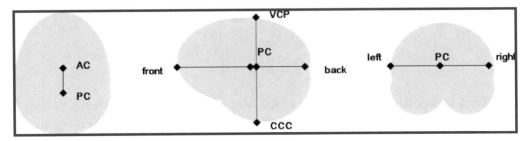

FIGURE 3.2: Anatomical landmarks as used for MNI transformation.

3.2.2 Second Iteration Step

In order to improve the quality of the template, template $A_{\text{templ},1}$ is used for the template-dependent transformation of the original data. For this transformation, a spherical coordinate frame is set up, and dilation factors along about 400 arrows (center at PC) are calculated (Fig. 3.3).

The mismatch between the intensities along an arrow of the volume of the data set to be fitted (via edge detection as well as via correlation) and the volume of the template $A_{\text{templ},1}$ is minimized according to the squared differences (χ^2). The dilation factors are then interpolated at the respective voxel position for each voxel to get the normalization on a voxelwise basis. Template 2 ($A_{\text{templ},2}$) is created by arithmetic averaging of the voxel intensities. That way, a modality-specific

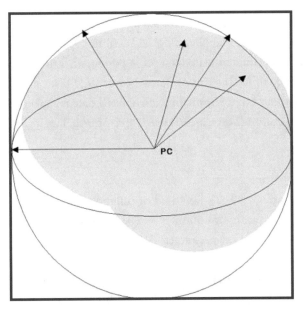

FIGURE 3.3: Template-dependent transformation – spherical coordinate frame with center at PC.

template is created. Single subject modality-specific data sets can be normalized according to MNI dimensions.

The MNI transformation for each individual data set is performed by minimizing the mismatch between intensity regions of the volume of the data set to be fitted (following the basic ideas of Ashburner and Friston [78]) and of the template $A_{\text{templ,2}}$ according to the squared differences (χ^2). Using the resulting transformation matrix, the other volumes of the modality are then to be transformed with the identical transformation parameters.

CHAPTER 4

Examples for Multimodal Imaging

To increase the understanding regarding the functional organization of the human brain, different aspects of the functional neuroanatomy must be considered. This includes, but is not limited to, the specific localization of eloquent areas, the type of processing in the brain, the time course of processing, and the interactions between the functions. The correlation of anatomy and function is one of the most important parameters in modern brain imaging analysis techniques. With this approach, different imaging techniques offer tradeoffs between spatial and temporal resolution [79]. In the following sections different approaches to combine information from different modalities are presented.

4.1 INTERMODAL MULTIMODALITY: MAGNETOENCEPHALOGRAPHY AND FUNCTIONAL MAGNETIC RESONANCE IMAGING

As we have seen in Section 2.2, MEG is a technique that provides a very high temporal resolution (i.e., on the order of 1 ms), mainly restricted only by the sampling frequency of the recording system [80]. Since magnetic fields are not distorted by conduction inhomogeneities in the head, MEG is a potentially powerful localization tool of brain function, providing spatial discrimination below 5 mm. However, MEG has a fundamental source identification problem, the so-called inverse problem: determining the active site in the brain from the magnetic field pattern recorded outside the skull. Thus, some a priori assumptions must be made about the source [63]. Usually, signal generators in the brain are described as current dipoles, which are physiologically reasonable models for relatively small active cortical areas.

Functional magnetic resonance imaging (fMRI) is a different approach to noninvasive analysis of the brain function, which has shown of widespread use over the last decade. The fMRI, based upon the Blood Oxygenation Level Dependent (BOLD) contrast, measures changes in the local concentration of deoxyhemoglobin by susceptibility-weighted MRI and can provide excellent spatial sampling in the range of 1 mm [81]. Therefore, it is obvious that the two measurement techniques, MEG and fMRI, detect signals with completely different physical origins. The localization of this origin is not expected to coincide for both techniques. The temporal resolution, due to the intrinsic inertia of blood flow changes, has a lower limit of about 100 ms. Moreover, the biological mechanisms underlying the BOLD contrast are complex because of many variables such as blood flow, blood volume, and subsequent oxygen change including factors like age of the subject [82].

These methodologies constrain and complement each other. When combined, they can therefore improve our interpretation of functional neural organization. Consecutively, it is useful to investigate brain function in a multimodal noninvasive way by combining different imaging techniques, thus making use of their individual advantages. Furthermore, the acquisition of individual volume-rendering morphological MRI data allows improved models of the head volume conductor. With particular respect to the inverse problem, one approach to analyzing MEG data is to impose anatomical and physiological constraints for neural electromagnetic source modeling derived from other imaging modalities such as fMRI to reduce the solution space. Combinations of MEG and fMRI in the same paradigms were described in several studies [5,2,3,4], both as a comparison of the localization of the results and an application of fMRI constraints to MEG data analysis. One example of a bimodal study using the latter approach is explained here in detail [83].

4.1.1   Results from Magnetoencephalography

The MEG measurements were performed on the 55-channel magnetometer system in Ulm, Germany [84]. The sensor system operated inside a magnetically shielded room [85] and consisted of a planar Dewar, containing a complex architectural structure with sensors distributed over two levels. The first level, i.e., the primary measurement plane, held 55 SQUID sensors. The sensing elements were integrated magnetometers with a square shape of 12.7 mm in diagonal. The sensors were uniformly distributed over the inner surface of the dewar, according to a hexagonal grid, covering a circular surface of about 230 mm in diameter. The measurement plane was 18 mm from the outer dewar bottom. Nineteen additional SQUIDs were mounted on the second level and were used as reference channels 90 mm from the measurement plane. Operating in the shielded environment, the system showed on all channels a white noise level better than $10 fT/\sqrt{Hz}$ at 10 Hz.

A software gradiometer setup was performed by subtracting online the background field sensed by selected reference channels from the signal of each primary channel. Three orthogonal electrocardiographic (ECG) leads for offline cancellation of the cardiac components, three bipolar EEG leads, and a respiration channel were used in addition. The recording sampling frequency was 8200 Hz. Signals were digitally filtered and decimated to a final sampling frequency of 1025 Hz and a bandwidth between DC and 250 Hz.

The standard measurement protocol of motor-related activity consisted of repeated flexion movements (FM) of the index finger of the right hand every 1.5 s. Five minutes of magnetic activity were recorded for the task. The self-paced movements were performed with a frequency of about 40 movements per min. The planar Dewar was positioned approximately 1 cm over the parietal cortex of the hemisphere contralateral to the movements (left hemisphere), covering the pericentral region. As a trigger signal for the finger movements, the surface myogram of the flexor digitorum muscle was recorded.

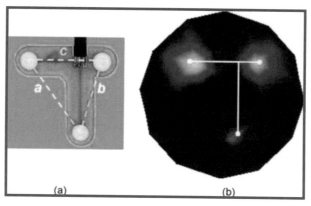

FIGURE 4.1: MEG recording determines the locations of the coils. (a) Schematic coil arrangement. (b) Magnetic field map reconstructed from a 15-s coil measurement with a 55-channel magnetometer system.

For the localization of the MEG marker coils, the coils were activated for a 15-s recording before the measurement started (Fig. 4.1).

4.1.2   Results from Functional Magnetic Resonance Imaging

The MRI and fMRI data were recorded with a 1.5 T scanner (Siemens, Magnetom, Erlangen, Germany) using echo planar imaging (EPI). For the acquisition of the volume-rendering MRI data set of the whole brain, a T_1-weighted magnetization-prepared rapid-acquisition gradient echo sequence was used (MP-RAGE, $T_R = 9.7$ ms, $T_E = 3.93$ ms, flip angle $15°$, matrix size 256×256 mm^2, voxel size $1.0 \times 1.0 \times 1.0$ mm^3), consisting of 160–200 sagittal partitions depending on the head size.

Functional data sets consisted of 130 recorded volumes in a block design changing between 10 volumes of rest (each 3 s) and 10 volumes of activation (each 3 s), 32 axial slices with 64×64 voxel each (voxel size $3.6 \times 3.6 \times 4.0$ mm^3, field of view (FOV) $230 \times 230 \times 128$ mm^3, $T_R = 50$ ms, $T_R = 2820$ ms).

The fMRI activation maps were overlaid on each individual 3-D brain. The respective T_1-weighted morphology covering the identical area as the functional data sets consisted of 32 slices with 256×256 voxels (voxel size $0.9 \times 0.9 \times 4.0$ mm^3, FOV $230 \times 230 \times 128$ mm^3, $T_E = 12$ ms, $T_E = 468$ ms). This data set was used for the coregistration of the functional localization results to the whole head recordings.

The whole head volume-rendering structural MRI data set serves, additionally, as the morphological basis for the superposition of the MEG results (see below). The true-scale imaging accuracy was tested for this pulse sequence with a maximum error of 1 mm [74].

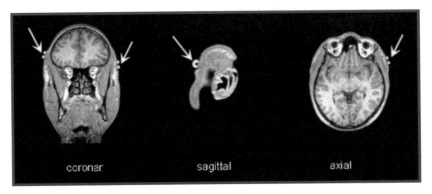

FIGURE 4.2: MRI marker detection. Hydrogel-based marker chips show up in MRI scans as ring structures (see arrows) and can be located manually by the operator.

4.1.3   Coregistering – Optical Marker Detection

For MSI, the combination of MEG data with the morphological data provided by the MRI is necessary. Coregistration may be based on distinct locations (marked by fiducials, electrodes or similar devices) or on surface matching algorithms using numerous intrinsic reference points. The adjustment of morphological MRI and MEG data in this example is accomplished by an optical localizer for the reasons outlined below. Optical systems can work with any well visible marker. For this reason, of course, the MRI markers and the MEG coils have to be prepared in a way that they produce striking optical images, e.g., color encoding. Combined fMRI/MEG recordings require different marker positions for the different modalities. These different marker positions are then transferred into one coordinate frame. Three MRI fiducial markers at suitable anatomical locations (Fig. 4.2) and three MEG markers (Fig. 4.3) were optically detected and coregistered by the optical coregistration system (OptiCoS) (Section 3.1) [86].

4.1.4   Multimodal Signal Processing and Results

The fMRI data during the motor activation paradigm were recorded according to the protocol described above. A detailed discussion of fMRI data processing and analysis algorithms is given in [87]. In that work, it has been shown by computer simulations that basically, deterministic analysis leads to an activation concept that is independent from the noise level, and vice versa, the statistical analysis leads to an activation concept where the activation is intended as deviation from noise and in such way that the activation becomes related to the noise level. The deterministic methods exclude true positive results with increasing noise and the statistical methods include false positive results with increasing noise. Nevertheless, a decision which of the analysis methods should be the method of choice for fMRI-analysis could not be made. In the example of data recorded from a right hand second finger movement, all the analysis methods listed here showed identical results (Fig. 4.4).

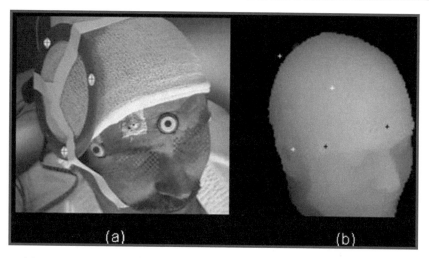

FIGURE 4.3: (a) Images of human head acquired by six cameras. Two marker types can be distinguished: MRI capsules marked with black dots and magnetic coils marked with light dots. (b) Artificial 3-D projection resulting from optical coregistration shows MEG markers as well as MRI markers.

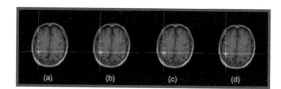

FIGURE 4.4: fMRI analysis of data recorded from a right hand second finger movement with different analysis methods (axial slice). (a) Deterministic correlation analysis. (b) Deterministic sinus wave analysis. (c) Statistical Student's t-test. (d) Statistical Kolmogorov–Smirnov test.

The functional regions in this course of experiments were identified by calculating the statistical test values [Eq. (2.32)].

In Fig. 4.5, functional activation in the primary motor cortex for FM is displayed. The T_1-weighted MRI data (covering the identical area as the fMRI data) are fitted to the T_1-weighted MP-RAGE data. Thus, the fMRI activation results can be superimposed onto the 3-D MRI data.

In addition, a segmented and triangulated surface is created from the MP-RAGE data. Using these surfaces as volume conductor models, the dipole localization can be performed (see below).

Epochs for averaging had to be selected from the MEG signal recordings. Fig. 4.6a shows examples for trigger signals from the specified FM myogram. A time-point in the trigger signal was defined automatically (occasionally having to be interactively corrected) for the trigger epochs that were defined by a similar selection process. Using these trigger time-points for each epoch, the series of signal epochs are selected for averaging. Fig. 4.6b–d show examples for averaged MEG signal epochs in different layout displays.

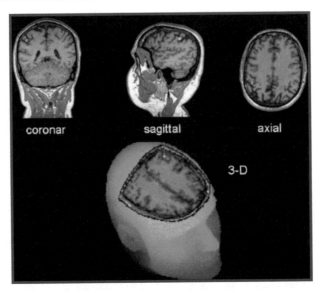

FIGURE 4.5: Functional activation in the left primary cortex (yellow/red) obtained by fMRI for a motor task (FM). The results are superimposed on the MP-RAGE.

Fig. 4.7 shows MEG source reconstruction results. To obtain the structural background for MEG source reconstruction, 3-D models from MRI data are generated and used as a morphological basis for the reconstruction. The triangulated surface is displayed in Fig. 4.7a. An ellipsoid that is fitted to the triangulated surface is shown in Fig. 4.7b. This ellipsoid can also act as volume conductor to simplify the source reconstruction.

Fig. 4.8a shows the correspondence between functional brain activation in the primary motor cortex identified by fMRI and the dipole localization obtained from MEG for a time-point 55 ms after myogram onset (Fig. 4.6a). The localization results for 20 subsequent time-points around the selected time-point revealed an average distance to the fMRI in the range of millimeters. To obtain interactivity between the modalities, the positions of fMRI active regions can be transferred to the MEG dipole localization procedure as seeds for the first guess or as possible dipole locations. The dipole localization result with the use of fMRI seeds (yellow) as first guess in the fitting procedure is shown in Fig. 4.8b.

It cannot be expected that the source of the magnetic field distribution for a certain time-point (resulting from postsynaptic potentials) exactly match with the activity revealed by the BOLD effect, which integrates over the entire block of movements, sometimes described as the "brain or vein" conflict. Localization differences in the range of several millimeters are therefore reasonable.

This accuracy is sufficient to use the regions identified as "active" in the fMRI analysis as possible sites of sources or as first guess for the fit for MEG source reconstruction.

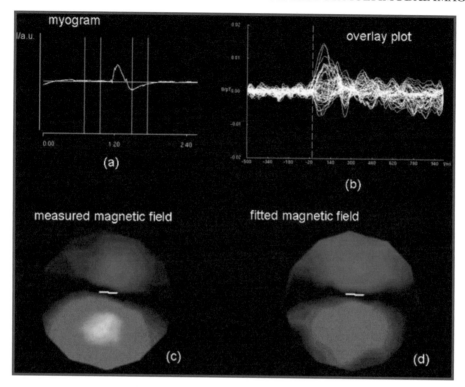

FIGURE 4.6: (a) Display of rectified electromyogram time course used as trigger for flexion movement producing MEG detectable brain activation. (b) Averaged MEG signal epochs of motor-evoked fields in overlay display, the dashed line corresponds to the trigger time-point in (a), the solid line marks the time-point of the dipole fit. (c) Measured magnetic field pattern for a defined time-point. (d) Fitted magnetic field pattern.

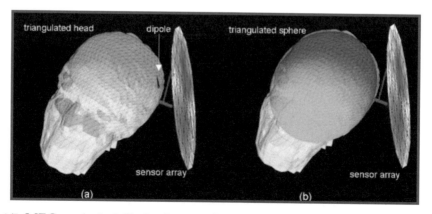

FIGURE 4.7: MEG results in 3-D visualization. (a) Triangulated head surface (about 1700 triangles) with dipole localization result displayed as a pin. (b) Idealized ellipsoid as an alternative conductor model.

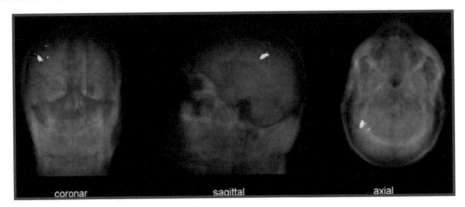

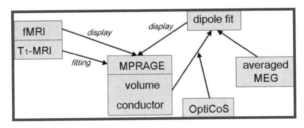

FIGURE 4.8: Functional activation in the left primary motor cortex (yellow) obtained by fMRI. The results are displayed on the morphological background of MRI 3-D recordings. The glass view represents the average over all respective slices. The single dipole localization is superimposed as a red dot.

FIGURE 4.9: Intermodal multimodality scheme.

To pursue synergetic interactions between the results of the two modalities, MEG and fMRI, it is essential to perform the combined analyses in one single software environment. The optical coregistration system (OptiCoS) provides the coordinate transformations and from hereon all analysis work is performed in one unique coordinate frame bound to the human head. All morphological and nonmorphologic information are available in one unique coordinate frame. For an overview, Fig. 4.9 illustrates major parts of this intermodal multimodality analysis concept. With the help of the MP-RAGE data and the optical marker detection by OptiCoS, the results of fMRI localization and dipole localization are overlaid on the identical individual MP-RAGE. Thus, comprehensive information of fMRI localization with MEG becomes feasible.

4.2   INTRAMODAL MULTIMODALITY: DIFFUSION TENSOR IMAGING AND MORPHOMETRY

As an example for intramodal multimodality, differences between normal subject groups and patients with certain disorders can be quantified. Comparison of groups of patients and healthy subjects can only be made after transformation of the single brain data into normalized stereotactic space. To explain the concurrence between diffusion tensor imaging (DTI) and morphometry, a study on patients with thinning of the corpus callosum (CC) was conducted. The subjects with CC thinning (which was present in the course of a neurodegenerative disease named hereditary spastic paraplegia (HSP) [88]) seemed particularly prone to this kind of analysis, since the CC is one of the cerebral structures/areas with the strongest accumulation of oriented fibers connecting the two hemispheres. Assuming HSP to be a white-matter disease, DTI should be able to detect axonal loss or at least damage of oriented white-matter regions [89].

4.2.1   The TIFT-Software

The following results have been created by the tensor imaging and fiber tracking (TIFT) software [90]. This software allows the scenario of a complete intramodal multimodality analysis. TIFT provides various quantification and visualization possibilities for the analysis of DTI, DWI, fMRI, and 3-D morphology data. TIFT provides several analysis methods, quantification methods via region of interest (ROI) analysis, stereotactic normalization techniques, group averaging, several interactive fiber tracking algorithms, and various 2-D and 3-D visualization options. The analysis methods can be applied to single subject data before and after MNI normalization as well as to the patient groups. The structure of the software aims at minimization of operator dependency providing analysis in a fast, comprehensive, and reproducible way.

The TIFT software provides new features in several respects:

- Stereotactic normalization techniques to the MNI space as implemented in TIFT are the prerequisites for group analysis. The TIFT software includes a new normalization technique based on scanner and sequence-specific templates and therefore can be adapted or newly created by every user depending on the requirements.

- Single subject data sets can be individually normalized and thus, FA mapping, ROI analysis, and FT can be performed in MNI space. The averaging of FA maps allows for the comparison of healthy subjects and patients with a similar pattern of white-matter changes, e.g., certain neurodegenerative diseases, at group level.

- ROI analyses offer the possibility to quantify diffusion properties in defined brain regions demonstrating the differences between patients and controls. This feature can be applied to individual data (before or after MNI deformation) or to group-averaged data. The TIFT-software can combine the ROI analysis with fiber tracking. First studies of such techniques have been reported by other authors [91,92]. This feature is part of a new comprehensive technology for the analysis of disturbed interconnectivity.

- In the field of FT, the TIFT-software provides several techniques and a set of parameters in which the operator has a wide choice for optimizing the FT for the needs of the specific brain region under inspection. The implemented normalization and averaging techniques allow for FT at group level for the first time.

- The fast and operator-independent implementation can be made use of in presurgical mapping of white matter. After MRI scanning the whole procedure including DICOM-conversion, motion correction, eddy-current correction, FA analysis, and FT takes no more than 30 min.

4.2.2 Study Protocol

DTI scanning protocols were performed on a 1.5 T scanner (Symphony, Siemens Medical, Erlangen, Germany). Six healthy controls (three men, three women, average age 32.7 ± 4.5 years) and six patients with CC thinning due to complicated HSP (three men, three women, average age 32.5 ± 12.1 years), as model data for degenerative diseases in the CC, underwent scanning.

The DTI study protocol consisted of 13 volumes (45 slices, 128×128 voxel, slice thickness 2.2 mm, in-plane voxel size 1.5 mm $\times$ 1.5 mm) representing 12 gradient directions and one scan with gradient 0 (b_0). T_E and T_R were 93 ms and 8000 ms, respectively; b was 800 s/mm^2; five scans were averaged. For the acquisition of the volume-rendering MRI data set of the whole brain used in morphometry (MP-RAGE), a T_1-weighted magnetization-prepared rapid-acquisition gradient echo sequence was used (MP-RAGE, $T_R = 9.7$ ms, $T_E = 3.93$ ms, flip angle 15°, matrix size 256×256 mm^2, voxel size $1.0 \times 0.96 \times 0.96$ mm^3), consisting of 160–200 sagittal partitions depending on the head size.

4.2.3 Template Creation

For the transformation into MNI space, the creation of study specific templates is necessary (see scheme in Fig. 4.10). The templates were created by the rules explained in Chapter 3.

The normalization of the MP-RAGE data sets is a straightforward procedure (for the created template see Fig. 4.11) whereas the normalization of DTI data sets is more sophisticated.

The normalization parameters are calculated for the b_0 volume and have to be applied consequently to all other volumes of the recording. The functional information also has to be transformed. The rotation parameters of the rigid brain transformation have to be stored so that the tensor $\vec{\vec{D}}$ of each voxel can be aligned by rotation to the main gradient directions when the MNI normalization of the DTI data sets is performed. Following the basic ideas of Alexander et al. [93], the dilation matrices are used for the alignment of the tensor $\vec{\vec{D}}$ of each voxel to the surrounding vox-

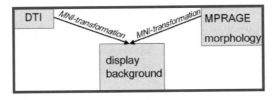

FIGURE 4.10: Intramodal multimodality scheme.

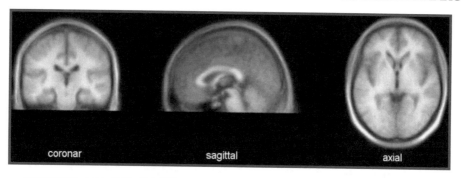

FIGURE 4.11: MP-RAGE template for normalization. Display focus is PC.

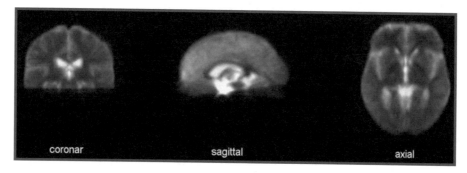

FIGURE 4.12: DTI template for normalization. Display focus is PC.

els resulting in a 3-D vector matrix containing three rotation angles at each voxel site. Strategies for spatial transformation of DTI data sets were also suggested by Ashburner et al. [78] and Park et al. [94,95].

Arithmetic averaging of all normalized single subject DTI data sets leads to an averaged DTI data set (Fig. 4.12). This approach is chosen in order to address two independent items. First, the voxel intensities of all subject's DTI data sets are arithmetically averaged. Second, the tensor fine-correction for each subject's data set (resulting from eddy current correction and normalization correction), i.e., a 3-D vector matrix for each data set, is arithmetically averaged. This averaged tensor fine-correction is applied to the calculation of the FA map of the averaged DTI data set.

In Fig. 4.13 a visual comparison between FA maps before and after MNI normalization is shown. ROI analyses have shown coincidence for the FA values before and after normalization [90].

4.2.4   Averaging for Different Subjects

Studies at group level may be important if the common clinical phenotype is supposed to be due to lesion of defined brain areas. Consequently, averaging of results for different subjects is necessary. Each individual brain has to be transferred into stereotactic space, and in a second step, the arithmetic averaging of the results is possible voxel by voxel. The averaging of the individual results can be performed in two ways:

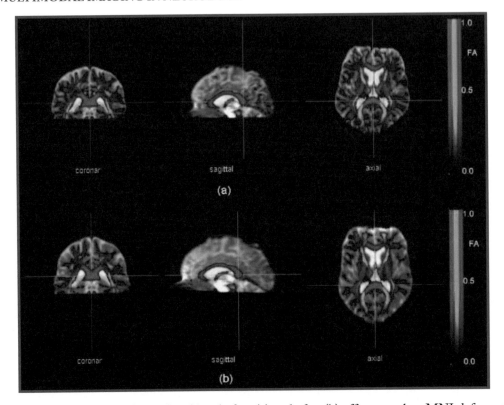

FIGURE 4.13: FA maps of a single subject before (a) and after (b) affine template MNI deformation. Display threshold for FA values is 0.2, background is the b_0 data set.

- For each single subject's data in MNI space, the FA map is calculated separately, and arithmetic averaging of the FA maps is performed afterward (Fig. 4.14a and b). Thus, a statistical test between the FA maps of the groups can be carried out.

- Each DTI data set is normalized and before FA mapping the whole DTI data sets are averaged. FA parameterization of these group-averaged DTI data leads to a group-specific FA map FAM2 (Fig. 4.15).

4.2.5   Examples for Fiber Tracking

From all data sets, an averaged DTI data set can be created by arithmetic averaging. Then, FT can be performed, taking the fine-corrections into account. Fig. 4.16a shows the FT for a healthy subject after MNI normalization (cf. Fig. 4.13). Fig. 4.16b shows FT performed on an averaged DTI data set, composed by arithmetic averaging of several MNI normalized data sets.

The application of the technique of DTI data averaging on the healthy subject group and the subject group with thinned CC enables detection of differences (Fig. 4.14). The thinning of the CC leads to reduced FA values and thus to restricted possibilities in FT (Fig. 4.17).

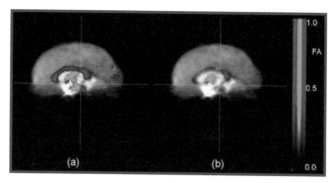

FIGURE 4.14: Arithmetic averaging of the FA maps after MNI normalization (FAM1). Display background is the averaged DTI data set (averaged b_0 volume), display threshold is 0.2, and background is the averaged b_0 data set. (a) Volunteers. (b) Patients with thinned CC.

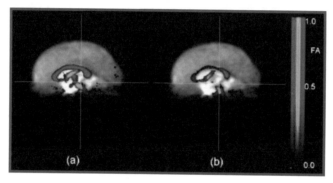

FIGURE 4.15: Arithmetic averaging of the FA maps after MNI normalization (FAM2). Display background is the averaged DTI data set (averaged b_0 volume), display threshold is 0.2, and background is the averaged b_0 data set. (a) Volunteers. (b) Patients with thinned CC.

The fiber tracks can be used as a skeleton mask for a selected statistics. The efforts of tract-based spatial statistics (TBSS) have been described by [62,96,97]. Here, a modified version named tractwise fractional anisotropy statistics (TFAS) is presented.

The resulting voxels from FT are used as skeleton. When FT is performed on averaged DTI data, each voxel that is crossed by a FT is defined as "active" for statistics. The underlying FA voxels contribute to the statistical t-test. The skeleton can be set up in different ways:

- For patients and volunteers skeletons are created independently and the FA voxel selection is performed independently for the volunteer and the patient data sets (skeleton 1).
- As a skeleton a combination of the skeletons from FT for volunteers and from FT for patients is used. In this way, the FA voxel selection is the same for patient data and volunteer data (skeleton 2).

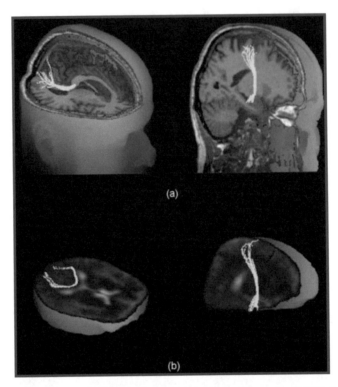

FIGURE 4.16: Fiber tracking results – 3-D views. (a) Healthy subject after MNI deformation displayed on the MP-RAGE as background: starting point in the dorsal corpus callosum (*left*) and starting point in the left internal capsule (*right*). (b) Averaged DTI data set from a group of healthy subjects: starting point in the dorsal corpus callosum (*left*) and starting point in the left internal capsule (*right*).

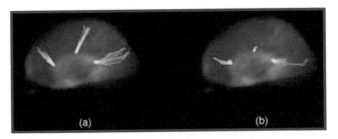

FIGURE 4.17: Examples for FT. Starting points were chosen in the CC. (a) Normal subject groups. (b) Patients with thinned CC. The images are glass-brain plots of the FTs overlaid on glass-brain plots of b_0 data.

Average FA and standard deviation are calculated by

$$F_{avg} = \frac{1}{N} \sum F_i, \quad \forall \text{voxel} \quad i \in \text{skeleton} \qquad (4.60)$$

$$F_{std} = \sqrt{\frac{1}{N-1} \sum (F_{avg} - F_i)^2}, \quad \forall \text{voxel} \quad i \in \text{skeleton} \qquad (4.61)$$

where N is the respective skeleton size, which usually is different for different skeletons. Then, TFAS uses F_{avg} and F_{std} for statistical t-test.

For both skeletons significant differences in TFAS analysis can be detected. Averaged FA values for healthy subjects are 0.43 ± 0.19 for skeleton 1 and 0.41 ± 0.19 for skeleton 2, respectively, whereas averaged FA values for patients with thinned CC are reduced to 0.25 ± 0.12 for skeleton 1 and to 0.21 ± 0.10 for skeleton 2, respectively.

4.2.6   Voxel-Based Morphometry

The so-called computational neuroanatomy is a methodology to characterize neuroanatomic configuration of different brains, encompassing voxel-based morphometry (VBM), which compares neuroanatomic differences on a voxel by voxel basis. Other techniques are deformation-based morphometry (DBM), which provides information about global differences in brain shape, and tensor-based morphometry (TBM), which provides information about local shape differences [98–100]. Especially VBM has shown to be a powerful tool for analyzing changes in gray or white matter volumes of the brain since its introduction in 1995 by use of an automated segmentation procedure and standardized parametric statistics in comparison with a normal data base within the Statistical Parametric Mapping (SPM) software (Wellcome Department of Imaging Neuroscience Group, London, UK; http://www.fil.ion.ucl.ac.uk/spm). VBM allows extracting information for each voxel separately from its gray-scale intensity.

For an extensive study of the VBM method with all details, the reader is referred to [101] and references therein. Other works of Ashburner et al. can help to complement the understanding [102,78,103]. A short description of the method is given in the following paragraphs:

The normalization procedure of the individual high-resolution brain data set (voxel size approx 1 mm^3) to a standard stereotactic space is the prerequisite. By nonlinear registration of brain images, a Jacobian matrix field is obtained in which each element is a tensor describing the relative positions of the neighboring elements. This is achieved by registering each of the images to the same template image by minimizing the residual sum of squared differences between them, using a 12 parameter affine transformation [102]. Thus, a Bayesian framework is used taking prior knowledge of the normal variability of brain size into account. In a second step global nonlinear shape differences are modeled by a linear combination of smooth spatial basis functions [78]. It has to be noted that this

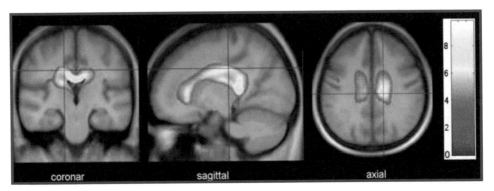

FIGURE 4.18: VBM results for patients with thinned CC versus normal subject groups. Displayed are *z*-score values on a standardized morphological background.

method of spatial normalization does not attempt to match every cortical feature exactly, but merely corrects for global brain shape differences.

After stereotactic transformation the images are partitioned into gray matter, white matter, cerebrospinal fluid and other background classes. It has to be noted that because the tissue classification is based on voxel intensities, image intensity nonuniformities have to be corrected prior to this segmentation procedure.

Next image processing steps are smoothing by convolving with an isotropic Gaussian kernel, and transformation of the local voxel concentrations (intensities) with the so-called logit function:

$$\text{logit}(I) = \frac{1}{2} \ln \frac{I}{1-I} \qquad (4.62)$$

After all these preprocessing steps, a statistical comparison on a voxel-by-voxel basis can be made. Fig. 4.18 shows an example for comparison of healthy subjects and patients with thinned CC calculated by the SPM software.

4.2.7   Intensity-Based 3-D MRI Analysis

This optimized VBM technique is rather complex. Therefore, in analogy to VBM, intensity-based 3D-MRI analysis (IBA) is used here to demonstrate the complementary results for IBA/VBM and DTI.

The MP-RAGE data sets for each single subject are normalized to the template shown in Fig. 4.11 using the techniques explained in Section 3.2. Arithmetic averaging of the normalized data sets for each group (patients and volunteers) leads to the data displayed in Fig. 4.19a and b.

In Fig. 4.20 Statistical analysis results of the IBA method comparing normal subject groups and patients with thinned CC are shown. Comparison with the VBM method (Fig. 4.18) shows coincidence.

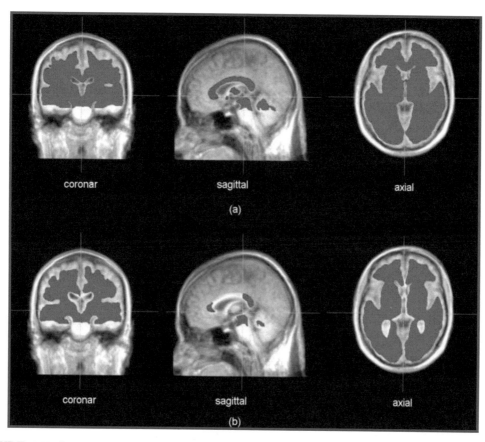

FIGURE 4.19: Intensity-based analysis (IBA) results, display threshold is 20% of maximum intensity, display background is the MP-RAGE template. (a) Averaged intensity image of volunteers. (b) Averaged intensity image of patients with thinned CC.

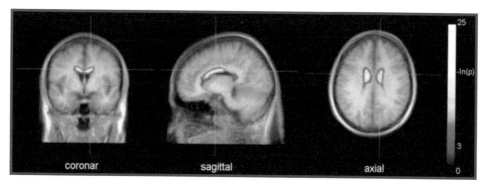

FIGURE 4.20: IBA results for patients with thinned CC versus normal subject groups (*t*-test analysis in white matter morphology). Display threshold is $p < 0.05$.

4.2.8   Complementary Intramodal Multimodality Results

IBA results and DTI results can provide complementary information displayed on the same MNI normalized background, making comparison feasible. In Fig. 4.21, results from two different MRI techniques are juxtaposed. Fig. 4.21a shows differences in white matter morphology between normal subject groups and patients with thinned CC extracted by IBA whereas Fig. 4.21b displays white matter diffusion anomalies extracted by FA analysis of DTI-weighted data sets.

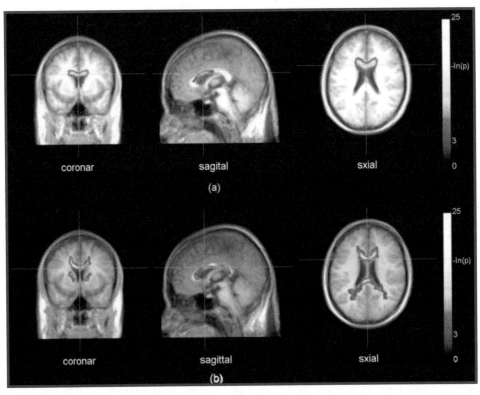

FIGURE 4.21: Complementary intramodal multimodality: (a) Differences in white matter morphology between normal subject groups and patients with thinned CC extracted by IBA. Display threshold difference is 40% of maximum intensity. (b) FA analysis of DTI-weighted data sets for patients with thinned CC versus normal subject groups (*t*-test analysis of FA maps). Display threshold is $p < 0.05$.

4.3   COMBINATION OF INTERMODAL AND INTRAMODAL MULTIMODALITIES

With the techniques described in Sections 1 and 2, the following comprehensive opportunities are suggested:

- IBA/VBM results might be used as seeds for FT. The starting points of FT might be localized adjacent to the IBA/VBM results and effects on FT might be topic of research. A powerful tool in this respect may be TFAS.

- fMRI functional results are not only complemented by their time development (MEG), but also can be combined to VBM results.

- fMRI active regions can act as seeds for FT. Thus, the FT starting points are localized nearby the fMRI active region. Interconnectivity of brain regions is thereby the major aim of research.

Of course, other neuroimaging techniques can complement the results explained in this chapter, e.g., multimodal imaging of PET, fMRI, and MEG [104], as additional applications of the same basic concepts.

CHAPTER 5

Clinical Aspects of Multimodal Imaging

Beyond general gain of information by multimodal neuroimaging, e.g., on connectivity of the brain [105], the future of multimodal imaging seems to be closely linked to its clinical applications. Two conceptual frameworks are of highest practical relevance: first, investigations with respect to understanding and diagnosis of neurological disorders with changes assessable by structural and/or functional neuroimaging techniques, in particular the large group of neurodegenerative diseases. One example, corpus callosum (CC) thinning due to the neurodegenerative disorder hereditary spastic paraparesis, has already been shown in Chapter 4. The second clinical application is the combination of different noninvasive neuroimaging tools and their cointegration into frameless stereotaxy/ neuronavigational systems after registration as an important element of presurgical diagnostics.

5.1 MULTIMODAL IMAGING IN THE RESEARCH OF NEUROLOGICAL (NEURODEGENERATIVE) DISEASES

Different MRI analysis techniques have been used in order to identify morphological signatures of brain diseases. In particular, the group of the neurodegenerative diseases has been extensively studied. Here, the computerized analysis techniques of structural MRI data (optimally acquired in a volume-rendering mode) are used that can capture the extraordinary morphological variability of the human brain, since these methods such as VBM use mathematical models sensitive to subtle changes in the size, position, shape, and tissue characteristics of brain structures affected by neurodegenerative diseases (cf. [106] for a review). However, beyond this volumetric or morphometric analysis of T_1-weighted MRI, various MRI applications in the sense of intramodal multimodality have proved useful both as diagnostic tools in individual patients and as instruments for the assessment of pathoanatomical and pathophysiological concepts at group level; at its simplest, this means a combination of structural MRI including volumetry and morphometry, DTI, and MRS. Because of this multimodal capability, MR has become the most versatile method for differential diagnosis in dementia with respect to the differential diagnosis and subcategorization and in order to monitor the progression and treatment effects [107]. Also in other important neurodegenerative disorders such as idiopathic Parkinson's disease and other Parkinsonian syndromes (cf. [108] for a review) and in motor neuron disorders (cf. [109] for a review), MRI applications investigating different aspects of the disorder-associated pathoanatomy have massively increased the clinicians' understanding over the recent years. In particular, in Parkinsonian syndromes, but also in other neurodegenerative disorders, the combination of MRI with (multiligand) nuclear medicine techniques is a valuable tool [110]. In multiple sclerosis, i.e., the most important chronic inflammatory disorder of the central nervous system, many different MRI techniques such as conventional MRI [111], DTI [112], fMRI [113], MRS [114], and a technique called magnetization transfer MRI [115] are important elements in diagnostics, differential diagnostics, and in improving our understanding of the lesions and the reparation procedures in the central nervous system. fMRI is still limited in its clinical applicability and has no relevant function in clinical diagnostics to date, but does enable the formulation of neurobiological hypotheses that can be tested clinically, and hence is well suited for testing classic clinical hypotheses about how the brain works so that understanding the mechanisms and sites of pathology might facilitate the development of new therapeutic strategies [116]. Also for MEG/MSI, the clinical use is limited to aspects such as the localization of pathological activity in association with neurological disorders such as certain epilepsy syndromes [117].

5.2   MULTIMODAL PRESURGICAL IMAGING

The potential of MSI for presurgical mapping of functionally important cortex has been extensively demonstrated [118,3,119,120,121,122]. MSI can be used for decision-making and planning of surgical treatment in brain lesions (in particular tumors) [123]. The most frequently used application is MSI with motor evoked fields and somatosensory evoked fields for mapping of the sensorimotor cortex in patients with lesions around the motor cortex [124,125,65]. Also, the determination of primary visual cortex by the investigation of visual evoked fields [126] as well as functional mapping of speech-related brain areas in individuals [4,127] were reported. MEG has a potential to directly detect and localize epilepsy-associated activity, i.e., MSI is used for presurgical evaluation in epilepsy surgery to identify the brain tissue generating epileptic activity [128,129,130,131].

With respect to the multimodality concept in the presurgical evaluation, a combination of functional neuroimaging techniques with basically different underlying neurophysiological correlates such as MEG/MSI and fMRI is most promising. Spatial comparisons between the localizational capability, in particular that of the sensorimotor cortex of MSI and fMRI, have been made by measuring the distances between MEG and fMRI activation sites after coregistration, both in healthy subjects [3] and in patients [4] showing different activation sites for the motor and sensory tasks within a range distance of 10 and 15 mm, probably as a reflection of the correlation between electrophysiologic and hemodynamic responses.

MSI results were obtained by the intraoperative cortical recording of somatosensory evoked potentials (so-called "phase reversal"), the localizations in the pre- and postcentral gyrus were verified in each case [125]. In a study with a similar technical approach using MEG and phase reversal technique in 30 patients with space-occupying lesions in or around the central region, MEG-based functional neuronavigation was also found to be practicable in finding a safe approach to tumors in or adjacent to the central region [132].

Both modalities MSI and fMRI are useful for the estimation of the sensorimotor cortex, and a combination of both methods has been used in order to avoid neurological deficits in certain cases [4]. Consecutively, the value of this integration of these two imaging modalities for spatio-temporal mapping of brain activity was strengthened in several publications [133] and led to the cointegration of both techniques into presurgical protocols [123], especially in patients receiving epilepsy surgery [134,135,136]. As an advanced combination of different modalities, combined data of DTI-based tractography fMRI and MEG have been used for the functional presurgical mapping of language areas [137]. It is useful to perform postprocessing of MEG and fMRI data in one software environment in that context to avoid potential methodologically based problems [79]. For this purpose of integration of fMRI, 3-D MRI, and MSI in one unique coordinate frame, the activation achieved by fMRI analysis can be used as possible sites of current sources for the MEG source reconstruction as an elegant approach for multimodality integration which was suggested by the "open magnetic and electric graphic analysis (OMEGA)" software [83] and which has been described in detail in Chapter 4. Further, the multimodal presurgical concept might be achieved by the combination of MEG/MSI and functional MRS [138]. With respect to the combination of different MRI techniques, quantitative DTI and MRS have been used to determine changes in the pyramidal tract adjacent to gliomas [139].

Regarding presurgical delineation of lesions, in particular epileptogenic lesions prior to epilepsy surgical procedures, voxel-based 3-D MRI analysis tools (e.g., VBM) have been developed towards the goal of automated observer-independent analysis [140,141,142,143]. Furthermore, as a consequent continuation of the multimodality approach in the sense of combined voxel-based mapping [144], a combination of ligand imaging using PET [145] and SPM-based MRI analysis was proposed. It could be demonstrated in a study using the approach of voxel-based MRI analysis and subtraction ictal SPECT coregistered to MRI (SISCOM) that multimodal assessment of malformations of cortical development may contribute to advanced observer-independent preoperative assessment of seizure origin and thus improve presurgical diagnostics in symptomatic epilepsy [146]. Although at the expense of the measurement of absolute quantitative values, the voxel-based comparison of MRI and PET enables the automated objective whole brain analysis, partial volume effects, and mixed tissue sampling in ligand images to be resolved, and disproportionate changes in function and structure to be identified (e.g., [147]). A detailed synopsis of multimodal imaging in presurgical diagnostics is given in [7].

5.3   ALTERNATIVE NEUROIMAGING TECHNIQUES

It has to be kept in mind that additional techniques such as (multichannel) EEG [148,149,150] exist, which is a major presurgical diagnostic element in epilepsy and in the localization prior to epilepsy surgery. However, EEG is rarely used for direct source coregistration with MRI and neuronavigation and was therefore not commented on, although in particular, the spatio-temporal relationship between MSI and EEG in epileptic focus localization has been subject of several studies [151,152], as well as the combination of EEG and fMRI [153] and, recently, the combination of EEG and MRS [154].

Furthermore, the technique of functional transcranial magnetic stimulation might provide an additional tool to add functional information (e.g., in the investigation of the virtual lesion).

5.4   ADVANTAGES AND LIMITATIONS

With respect to the cost effectiveness and the burden of often time-consuming data acquisition procedures in critically ill patients, multimodality imaging does not necessarily mean "the more, the better". Some options such as techniques for the localization of eloquent cortex (fMRI or MSI, respectively) and techniques for the imaging of white matter fiber bundles (DTI) are complementary. For the methods of imaging cortical areas, it has to be held that fMRI can be performed on most regular clinical MRI scanners, whereas MSI, although it seems to be a promising technique, is still limited because of the low distribution of MEG devices which will most probably not increase much over the coming years. Nuclear medicine techniques are valuable tools in the delineation of cerebral neoplasms or other lesions; in particular as multiligand imaging, these nuclear medicine techniques may be complemented by MRS in neurological diseases, especially neurodegenerative disorders.

If different neuroimaging modalities are used, the user has to be aware that the fusion of the results of these techniques is possible after coregistration, but that equivalent localizations in the ideal neuroanatomical sense are very difficult to achieve and sometimes impossible because of different neurophysiological substrates, acquisition methods, and spatial resolution.

5.5   FUTURE

The future of multimodal imaging should be seen in connection to its clinical applicability. Prerequisites for use of multimodal imaging in a clinical setting are listed as follows:

- Measurement devices should be available and measurement time should be comparatively short.
- Optimally, all analyses should be performed in one single software environment.
- Short computation time, i.e., optimal results should be produced within minutes after finishing the last measurement. For example, computation time on a high performance PC is about 1.0 s per volume for fMRI motion correction, 0.8 s for functional analysis and 0.2 s for a dipole fit [83].
- Results should be gained in a data postprocessing as operator-independent as possible, i.e., the software is running automatically with as few observer interactions as possible.

REFERENCES

[1] J. G. Beaumont, *Introduction to Neuropsychology*. New York: The Guilford Press; 1983.

[2] J. S. George, C. J. Aine, J. C. Mosher, D. M. Schmidt, D. M. Ranken, H. A. Schlitt, C. C. Wood, J. D. Lewine, J. A. Sanders, and J. W. Belliveau, "Mapping function in the human brain with magnetoencephalography, anatomical magnetic resonance imaging, and functional magnetic resonance imaging," *Clin. Neurophysiol.*, vol. 12, no. 5, pp. 406–431, 1995, doi:10.1097/00004691-199509010-00002.

[3] C. Stippich, P. Freitag, J. Kassubek, P. Sörös, K. Kamada, H. Kober, K. Scheffler, R. Hopfengärtner, D. Bilecen, E. W. Radü, and J. B. Vieth, "Motor, somatosensory and auditory cortex localization by fMRI and MEG," *NeuroReport*, vol. 9, no. 9, pp. 1953–1957, 1998, doi:10.1097/00001756-199806220-00007.

[4] H. Kober, C. Nimsky, M. Moller, P. Hastreiter, R. Fahlbusch, and O. Ganslandt, "Correlation of sensorimotor activation with functional magnetic resonance imaging and magnetoencephalography in presurgical functional imaging: A spatial analysis," *NeuroImage*, vol. 14, no. 5, pp. 1214–1228, 2001, doi:10.1006/nimg.2001.0909.

[5] N. Fujimaki, T. Hayakawa, M. Nielsen, T. R. Knosche, and S. Miyauchi, "An fMRI-constrained MEG source analysis with procedures for dividing and grouping activation," *NeuroImage*, vol. 17, no. 1, pp. 324–343, 2002, doi:10.1006/nimg.2002.1160.

[6] S. Thees, F. Blankenburg, B. Taskin, G. Curio, and A. Villringer, "Dipole source localization and fMRI of simultaneously recorded data applied to somatosensory categorization," *NeuroImage*, vol. 18, pp. 707–719, 2003, doi:10.1016/S1053-8119(02)00054-X.

[7] J. Kassubek, and F. D. Juengling, "Multimodality functional neuroimaging," in Stippich C, Ed. *Clinical Functional MRI – Presurgical Functional Neuroimaging*. Berlin/Heidelberg/New York: Springer; 2007, pp. 191–210.

[8] E. Fukushima and B. W. Roeder, *Experimental Pulse NMR – A Nuts and Bolts Approach*, 10th ed. Cambridge, MA: Perseus Publishing; 1986.

[9] P. T. Callaghan, *Principles of Nuclear Magnetic Resonance Microscopy*. Oxford: Clarendon Press; 1994.

[10] R. Kimmich, *NMR Tomography, Diffusometry, Relaxometry*. Berlin: Springer; 1997.

[11] S. Stapf, and S. Han, *NMR Imaging in Chemical Engineering*. Weinheim, Germany: Wiley; 2006.

[12] F. Bloch, W. W. Hansen, and M. Packard, "The nuclear induction experiment," *Phys. Rev.*, vol. 70, pp. 474–485, 1946, doi:10.1103/PhysRev.70.474.

[13] A. Haase, J. Frahm, D. Matthaei, W. Hänicke, and K. D. Merbboldt, "FLASH imaging, rapid NMR imaging using low flip angle pulses," *Magn. Reson.*, vol. 67, pp. 258–266, 1986, doi:10.1016/0022-2364(86)90433-6.

[14] A. Mishra, Y. Lu, J. Meng, A. W. Anderson, and Z. Ding Z, "Unified framework for anisotropic interpolation and smoothing of diffusion tensor images," *NeuroImage*, vol. 31, no. 4, pp. 1525–1535, 2006, doi:10.1016/j.neuroimage.2006.02.031.

[15] R. S. Frackowiak, K. J. Friston, C. D. Frith, R. J. Dolan, and J. C. Mazziotta, *Human Brain Function*. New York: Academic Press; 1997.

[16] A. W. Toga, and J. C. Mazziotta, *Brain Mapping – The Methods*. San Diego: Academic Press; 1996.

[17] D. Le Bihan, *Diffusion and Perfusion Magnetic Resonance Imaging – Applications to Functional MRI*. New York: Raven Press; 1995.

[18] S. Ogawa, T. M. Lee, A. R. Kay, and D. W. Tank, "Brain magnetic resonance imaging with contrast dependent on blood oxygenation," *Proc. Natl. Acad. Sci. USA*, vol. 87, no. 24, pp. 9868–9872, 1990, doi:10.1073/pnas.87.24.9868.

[19] S. Ogawa, T. M. Lee, A. S. Nayak, and P. Glynn, "Oxygenation-sensitive contrast in magnetic resonance image of rodent brain at high magnetic fields," *Magn. Reson. Med.*, vol. 14, no. 1, pp. 68–78, 1990, doi:10.1002/mrm.1910140108.

[20] P. C. van Zijl, S. M. Eleff, J. A. Ulatowski, J. M. Oja, A. M. Ulug, R. J. Traystman, and R. A. Kauppinen, "Quantitative assessment of blood flow, blood volume and blood oxygenation effects in functional magnetic resonance imaging," *Nat. Med.* vol. 42, no. 2, pp. 159–167, 1998, doi:10.1038/nm0298-159.

[21] S. Ogawa, T. M. Lee, and B. Barrere, "The sensitivity of magnetic resonance image signals of a rat brain to changes in the cerebral venous blood oxygenation," *Magn. Reson. Med.*, vol. 29, no. 2, pp. 205–210, 1993, doi:10.1002/mrm.1910290208.

[22] K. K. Kwong, "Functional MRI with echo planar imaging," *Mag. Reson. Quarterly*, vol. 11, pp. 1–20, 1995.

[23] R. Turner, "Functional mapping of the human brain with MRI," *Sem. Neurosci.*, vol. 7, pp. 179–194, 1995.

[24] J. V. Hajnal, R. Myers, A. Oatridge, J. E. Schwieso, I. R. Young, and G. M. Bydder, "Artifacts due to stimulus correlated motion in functional imaging of the brain," *Magn. Reson. Med.*, vol. 31, no. 3, pp. 283–291, 1994, doi:10.1002/mrm.1910310307.

[25] R. P. Woods, S. R. Cherry, and J. C. Mazziotta, "Rapid automated algorithm for aligning and reslicing PET images," *Comput. Assist. Tomogr.* vol. 16, no. 4, pp. 620–633, 1992, doi:10.1097/00004728-199207000-00024.

[26] K. J. Friston, J. Ashburner, C. D. Frith, J. B. Poline, J. D. Heather, and R. S. Frackowiak, "Spatial registration and normalization of images," *Hum. Brain Mapp.*, vol. 3, no. 3, pp. 165–189, 1995, doi:10.1002/hbm.460030303.

[27] W. Andrä, H. Nowak, *Magnetism in Medicine – A Handbook*. Wiley-VCH: Berlin; 1998.

[28] P. A. Bandettini, A. Jesmanowicz, E. C. Wong, and J. S. Hyde, "Processing strategies for time-course data sets in functional MRI of the human brain," *Magn. Reson. Med.*, vol. 30, no. 2, pp. 161–173, 1993, doi:10.1002/mrm.1910300204.

[29] K. Kuppusamy, W. Lin, and M. Haacke, "Statistical assessment of crosscorrelation and variance methods and the importance of electrocardiogram gating in functional magnetic resonance imaging," *Magn. Reson. Imaging*, vol. 15, pp. 169–181, 1997, doi:10.1016/S0730-725X(96)00338-4.

[30] P. Filzmoser, R. Baumgartner, and E. Moser, "A hierarchical clustering method for analyzing functional MR images," *Magn. Res. Imaging*, vol. 17, pp. 817–826, 1999, doi:10.1016/S0730-725X(99)00014-4.

[31] X. Ding, J. Tkach, P. Ruggieri, and T. Masaryk, "Analysis of time-course functional MRI data with clustering method without use of reference signal," *Proc. Int. Soc. Magn. Reson.*, pp. 630–635, 1994.

[32] W. H. Press, S. A. Teukolsky, W. T. Vetterling, and B. P. Flannery, *Numerical Recipes in C.* New York: Cambridge University Press; 1998.

[33] K. J. Friston, C. D. Frith, R. S. Frackowiak and R. Turner, "Characterizing dynamic brain responses with fMRI: A multivariate approach," *NeuroImage*, vol. 2, no. 2, pp. 166–172, 1995, doi:10.1006/nimg.1995.1019.

[34] K. J. Friston, C. D. Frith, R. Turner and R. S. Frackowiak, "Characterizing evoked hemodynamics with Fmri," *NeuroImage*, vol. 2, no. 2, pp. 157–165, 1995, doi:10.1006/nimg.1995.1018.

[35] K. J. Friston, A. P. Holmes, J. B. Poline, P. J. Grasby, S. C. Williams, R. S. Frackowiak, and R. Turner, "Analysis of fMRI time-series revisited," *NeuroImage*, vol. 2, no. 1, pp. 45–53, 1995, doi:10.1006/nimg.1995.1007.

[36] K. J. Friston, A. Holmes, J. B. Poline, C. J. Price, and C. D. Frith, "Detecting activations in PET and fMRI: Levels of inference and power," *NeuroImage*, vol. 4, pp. 223–235, 1996, doi:10.1006/nimg.1996.0074.

[37] K. J. Friston, S. Williams, R. Howard, R. S. Frackowiak, and R. Turner, "Movement-related effects in fMRI time-series," *Magn. Reson. Med.*, vol. 35, no. 3, pp. 346–355, 1996.

[38] K. J. Friston, C. D. Frith, P. Fletcher, P. F. Liddle, and R. S. Frackowiak "Functional topography: Multidimensional scaling and functional connectivity in the brain," *Cereb. Cortex*, vol. 6, no. 2, pp. 156–164, 1996, doi:10.1093/cercor/6.2.156.

[39] K. J. Friston, O. Josephs, G. Rees and R. Turner, "Nonlinear event-related responses in Fmri," *Magn. Reson. Med.*, vol. 39, no. 1, pp. 41–52, 1998, doi:10.1002/mrm.1910390109.

[40] D. Le Bihan, F. Breton, D. Lallemand, M. L. Aubin, J. Vignaud, and M. Laval-Jeantet, "Separation of diffusion and perfusionin intravoxel incoherent motion MR imaging," *Radiology*, vol. 168, pp. 497–505, 1988.

[41] E. R. Melhem, S. Mori, G. Mukundan, M. A. Kraut, M. G. Pomper, and P. C. van Zijl, "Diffusion tensor MR imaging of the brain and white matter tractography," *Am. J. Radiol.*, vol. 178, pp. 3–16, 2002.

[42] P. J Basser, J. Mattiello, and D. LeBihan, "Estimation of the effective self-diffusion tensor from the NMR spin echo," *Magn. Reson.*, vol. 103, no. 3, pp. 247–254, 1994, doi:10.1006/jmrb.1994.1037.

[43] P. J. Basser, J. Mattiello, and D. LeBihan, "MR diffusion tensor spectroscopy and imaging," *Biophysics*, vol. 66, no. 1, pp. 259–267, 1994.

[44] A. L. Tievsky, T. Ptak, and J. Farkas, "Investigation of apparent diffusion coefficient and diffusion tensor anisotropy in acute and chronic multiple sclerosis lesions," *Neuroradiology*, vol. 20, pp. 1491–1499, 1999.

[45] K. Yamada, O. Kizu, H. Ito, T. Kubota, W. Akada, M. Goto, A. Takada, J. Konishi, H. Sasajima, K. Mineura, S. Mori, and T. Nishimura, "Tractography for arteriovenous malformations near the sensorimotor cortices," *AJNR Neuroradiol.*, vol. 26, no. 3, pp. 598–602, 2005.

[46] S.-K. Lee, D. I. Kim, J. Kim, D. J. Kim, H. D. Kim, D. S. Kim, and S. Mori, "Diffusion-tensor MR imaging and fiber tractography: a new method of describing aberrant fiber connections in developmental CNS anomalies," *Radiographics*, vol. 25, pp. 53–65, 2005, doi:10.1148/rg.251045085.

[47] C. R. Tench, P. S. Morgan, M. Wilson, and L. D. Blumhardt, "White matter mapping using diffusion tensor MRI," *Magn. Reson. Med.*, vol. 47, pp. 967–972, 2002, doi:10.1002/mrm.10144.

[48] T. L. Chenevert, J. A. Brunberg, and J. G. Pipe, "Anisotropic diffusion in human white matter: demonstration with MR techniques in vivo," *Radiology*, vol. 177, no. 2, pp. 401–405, 1990.

[49] R. Turner, D. Le Bihan, J. Maier, R. Vavrek, L. K. Hedges, and J. Pekar, "Echo-planar imaging of intravoxel incoherent motion," *Radiology*, vol. 177, no. 2, pp. 407–414, 1990.

[50] E. O. Stejskal, and J. F. Tanner, "Spin diffusion measurements: spin echos in the presence of a time-dependent field gradient," *Chem. Phys.*, vol. 42, pp. 288–292, 1965, doi:10.1063/1.1695690.

[51] C. Pierpaoli, P. Jezzard, P. J. Basser, A. Barnett, and G. Di Chiro, "Diffusion tensor MR imaging of the human brain," *Radiology*, vol. 201, pp. 637–648, 1996.

[52] M. Neeman, J. P. Freyer, and L. O. Sillerud, "Pulsed-gradient spin-echo studies in NMR imaging. Effects of the imaging gradients on the determination of diffusion coefficients," *J. Magn. Reson.*, vol. 90, pp. 303–312, 1990, doi:10.1016/0022-2364(90)90136-W.

[53] D. Le Bihan, J.-F. Mangin, C. Poupon, C. A. Clark, S. Pappata, N. Molko, and H. Chabriat, "Diffusion tensor imaging: concepts and applications," *Magn. Reson.*, vol. 13, pp. 534–546, 2001, doi:10.1002/jmri.1076.

[54] P. J. Basser, and D. K. Jones, "Diffusion-tensor MRI: theory, experimental design and data analysis—a technical review," *NMR Biomed.*, vol. 15, no. 7–8, pp. 456–467, 2002, doi:10.1002/nbm.783.

[55] Y. Shen, D. J. Larkman, S. Counsell, I. M. Pu, D. Edwards, and J. V. Hajnal, "Correction of high-order eddy current induced geometric distortion in diffusion-weighted echo-planar images," *Magn. Reson. Med.* vol. 52, no. 5, pp. 1184–1189, 2004, doi:10.1002/mrm.20267.

[56] B. Herrnberger, "Diffusionstensor-Magnetresonanztomographie," *Nervenheilkunde*, vol. 23, pp. 50–59, 2004.

[57] P. J. Basser, S. Pajevic, C. Pierpaoli, J. Duda, and A. Aldroubi, "In vivo fiber tractography using DT-MRI data," *Magn. Reson. Med.*, vol. 44, no. 4, pp. 625–632, 2000, doi:10.1002/1522-2594(200010)44:4<625::AID-MRM17>3.0.CO;2-O.

[58] T. E. Conturo, N. F. Lori, T. S. Cull, E. Akbudak, A. Z. Snyder, J. S. Shimony, R. C. McKinstry, H. Burton, and M. E. Raichle, "Tracking neuronal fiber pathways in the living human brain," *Proc. Natl. Acad. Sci. USA*, vol. 96, no. 18, pp. 10422–10427, 1999, doi:10.1073/pnas.96.18.10422.

[59] S. Mori, B. Cain, V. P. Chacko, and P. C. van Zijl, "Three dimensional tracking of axonal projections in the brain by magnetic resonance imaging," *Ann. Neurol.*, vol. 45, pp. 265–269, 1999, doi:10.1002/1531-8249(199902)45:2<265::AID-ANA21>3.0.CO;2-3.

[60] M. Lazar, D. M. Weinstein, J. S. Tsuruda, K. M. Hasan, K. Arfanakis, M. E. Meyerand, B. Badie, H. A. Rowley, V. Haughton, A. Field, and A. L. Alexander, "White matter tractography using diffusion tensor deflection," *Hum. Brain Mapp.* vol. 18, no. 4, pp. 306–321, 2003, doi:10.1002/hbm.10102.

[61] D. K. Jones, A. Simmons, S. C. R. Williams, and M. A. Horsfield, "Noninvasive assessment of axonal fiber connectivity in the human brain via diffusion tensor MRI," *Magn. Reson. Med.*, vol. 42, pp. 37–41, 1999, doi:10.1002/(SICI)1522-2594(199907)42:1<37::AID-MRM7>3.0.CO;2-O.

[62] T. E. Behrens, H. J. Berg, S. Jbabdi, M. F. Rushworth, and M. W. Woolrich, "Probabilistic diffusion tractography with multiple fibre orientations: What can we gain?" *NeuroImage*, vol. 34, no. 1, pp. 144–155, 2007, doi:10.1016/j.neuroimage.2006.09.018.

[63] M. Hämäläinen, R. Hari, R. J. Ilmoniemi, J. Knuutila, and O. V. Lounasmaa, "Magnetoencephalography – theory, instrumentation, and applications to noninvasive

studies of the working human brain," *Rev. Mod. Phys.*, vol. 65, pp. 413–497, 1993, doi:10.1103/RevModPhys.65.413.

[64] K. J. Friston, K. M. Stephan, J. D. Heather, C. D. Frith, A. A. Ioannides, L. C. Liu, M. D. Rugg, J. Vieth, H. Keber, K. Hunter, and R. S. Frackowiak, "A multivariate analysis of evoked responses in EEG and MEG data," *NeuroImage*, vol. 3, pp. 167–174, 1996, doi:10.1006/nimg.1996.0018.

[65] O. Ganslandt, R. Fahlbusch, C. Nimsky, H. Kober, M. Moller, R. Steinmeier, J. Romstock, and J. Vieth, "Functional neuronavigation with magnetoencephalography: Outcome in 50 patients with lesions around the motor cortex," *Neurosurgery*, vol. 91, pp. 73–79, 1999.

[66] J. Kassubek, C. Stippich, P. Sörös, O. Ganslandt, K. Kamada, R. Hopfengärtner, H. Kober, R. Steinmeierv, and J. Vieth, "A motor field source localization protocol using magnetoencephalography," *Biomedizinische Technik.*, vol. 41, Suppl 1, pp. 334–335, 1996.

[67] M. Hämäläinen, and J. Sarvas, "Realistic conductivity geometry model of the human head for interpretation of neuromagnetic data," *IEEE Trans. Biomed. Eng.*, vol. 36, pp. 165–171, 1989, doi:10.1109/10.16463.

[68] D. Geselowitz, "On the magnetic field generated outside an inhomogeneous volume conductor by internal current sources," *IEEE Trans. Magn.*, vol. 6, pp. 346–347, 1970, doi:10.1109/TMAG.1970.1066765.

[69] S. Tissari, J. Rahola, "Error analysis of Galerkin method to solve the forward problem in MEG using the boundary element method," *Comput. Methods Programs Biomed.*, vol. 72, no. 3, pp. 209–222, 2003, doi:10.1016/S0169-2607(02)00144-X.

[70] A. S. Ferguson, X. Zhang and G. Stroink, "A complete linear discretization for calculating the magnetic field using the boundary element method," *IEEE Trans. Biomed. Eng.*, vol. 41, pp. 455–460, 1994, doi:10.1109/10.293220.

[71] J. de Munck, "A Linear discretization of the volume conductor boundary integral equation using analytically integrated elements," *IEEE Trans. Biomed. Eng.*, vol. 39, pp. 986–990, 1992, doi:10.1109/10.256433.

[72] C. Del Gratta, S. N. Erné, and J. Edrich, "A Linear iterative algorithm for dipole localization," in C. Baumgartner, L. Deecke, G. Stroink, and S. J. Williamson, Eds. *Biomagnetism: Fundamental Research and Clinical Applications*. Amsterdam: Elsevier Science; 1995, pp. 313–316.

[73] Polhemus, *3Spacereg Fasttrak User's Manual*. Colchester, Vermont, 1993.

[74] H. Lindenthal, H.-P. Müller, A. Pasquarelli, and S. N. Erné, "Versatile optical co-registration system for multimodal marker localization," *Biomed. Tech. (Berl)*, vol. 48, no. 2, pp. 272–274, 2004.

[75] J. Talairach, and P. Tournoux, *Coplanar Stereotaxic Atlas of the Human Brain*. New York: Thieme Medical; 1988.

[76]   D. L. Collins, P. Neeli, T. M. Peters, and A. C. Evans, "Automatic 3D intersubject registration of MR volumetric data in standardized Talairach space," *Comput. Assist. Tomogr.*, vol. 18, no. 2, pp. 192–205, 1994.

[77]   M. Brett, I. S. Johnsrude, and A. M. Owen, "The problem of functional localization in the human brain," *Nat. Rev. Neurosci.*, vol. 3, no. 3, pp. 243–249, 2002, doi:10.1038/nrn756.

[78]   J. Ashburner, and K. J. Friston, "Nonlinear spatial normalization using basis functions," *Hum Brain Mapp.*, vol. 7, pp. 254–266, 1999, doi:10.1002/(SICI)1097-0193(1999)7:4<254::AID-HBM4>3.0.CO;2-G.

[79]   A. M. Dale, A. K. Liu, B. R. Fischl, R. L. Buckner, J. W. Belliveau, J. D. Lewine, and E. Halgren, "Dynamic statistical parametric mapping: Combining fMRI and MEG for high-resolution imaging of cortical activity," *Neuron*, vol. 26, no. 1, pp. 55–67, 2000, doi:10.1016/S0896-6273(00)81138-1.

[80]   O. V. Lounasmaa, M. Hämäläinen, R. Hari and R. Salmelin, "Information processing in the human brain: magnetoencephalographic approach," *Proc. Natl. Acad. Sci. USA*, vol. 93, no. 17, pp. 8809–8815, 1996, doi:10.1073/pnas.93.17.8809.

[81]   K. K. Kwong, J. W. Belliveau, D. A. Chesler, I. E. Goldberg, R. M. Weisskoff, B. P. Poncelet, D. N. Kennedy, B. E. Hoppel, M. S. Cohen, R. Turner, et al., "Dynamic magnetic resonance imaging of human brain activity during primary sensory stimulation," *Proc. Natl. Acad. Sci. USA*, 1992, vol. 89, pp. 5675–5679, doi:10.1073/pnas.89.12.5675.

[82]   A. Riecker, W. Grodd, U. Klose, J. B. Schulz, K. Gröschel, M. Erb, H. Ackermann, and A. Kastrup, "Relation between regional functional MRI activation and vscular reactivity to carbon dioxide during normal aging," *Cereb. Blood Flow Metab.*, vol. 23, no. 5, pp. 565–573, 2003, doi:10.1097/01.WCB.0000056063.25434.04.

[83]   H-P Müller, I. de Cesaris, M. de Melis, L. Marzetti, A. Pasquarelli S. N. Erné, A. C. Ludolph, and J. Kassubek, "Open magnetic and electric graphic analysis: comprehensive magnetoencephalographic and functional magnetic resonance imaging in one single software environment," *IEEE Eng. Med. Biol. Mag.*, vol. 24, no. 3, pp. 109–116, 2005.

[84]   S. N. Erné, A. Pasquarelli, H. Kammrath, S. Della Penna, K. Torquati, V. Pizzella, R. Rossi, C. Granata, and M. Russo, *Argos 55—The New MCG System in Ulm,*" in T. Yoshimoto, M. Kotani, S. Kuriki, H. Karibe, N. Nakasato, Eds. *Recent Advances in Biomagnetism. BIOMAG'98*, pp. 27–30, 1998.

[85]   A. Pasquarelli, H. Kammrath, U. Tenner, and S. N. Erné, "The New Ulm Magnetically Shielded Room," in T. Yoshimoto, M. Kotani, S. Kuriki, H. Karibe, and N. Nakasato, Eds. *Recent Advances in Biomagnetism: BIOMAG'98*, 1998, pp. 55–58.

[86]   H. Lindenthal, H.-P. Müller, A. Pasquarelli, and S. N. Erné, "OptiCoS – a new optical co-registration system for multimodal integration," in *Proc. NFSI* 2001, Innsbruck, Austria, 2001, pp. 127–129.

[87] H.-P. Müller, E. Kraft, A. C. Ludolph, and S. N. Erné, "New analysis methods in fMRI analysis: hierarchical cluster analysis for improved signal-to-noise ratio compared to standard techniques," *IEEE Eng. Med. Biol. Mag.*, vol. 21, no. 5, pp. 134–142, 2002.

[88] J. Kassubek, F. D. Juengling, A. Baumgartner, A. C. Ludolph, and A.-D. Sperfeld, "Different regional brain volume loss in pure and complicated hereditary spastic paraparesis: a voxel-based morphometry study, Amyotrophic Leteral Sclerosis 2007," in press.

[89] S. Dreha-Kulaczewski, P. Dechent, G. Helms, J. Frahm, J. Gärtnerand K. Brockmann, "Cerebral metabolic and structural alterations in hereditary spastic paraplegia with thin corpus callosum assessed by MRS and DTI," *Neuroradiology*, vol. 48, no. 12, pp. 893–898, 2006, doi:10.1007/s00234-006-0148-2.

[90] H-P Müller, A. Unrath, A. C. Ludolph AC, and J. Kassubek, "Preservation of diffusion tensor properties during spatial normalization by use of tensor imaging and fiber tracking on a normal brain database," *Phys. Med. Biol.*, vol. 52, no. 6, pp. N99–N109, 2007, doi:10.1088/0031-9155/52/6/N01.

[91] T. E. Behrens, H. Johansen-Berg, M. W. Woolrich, S. M. Smith, C. A. Wheeler-Kingshott, P. A. Boulby, G. J. Barker, E. L. Sillery, K. Sheehan, O. Ciccarelli, A. J. Thompson, J. M. Brady, and P. M. Matthews, "Noninvasive mapping of connections between human thalamus and cortex using diffusion imaging," *Nat. Neurosci.*, vol. 6, no. 7, pp. 750–757, 2003, doi:10.1038/nn1075.

[92] R. A. Kanaan, S. S. Shergill, G. J. Barker, M. Catani, V. W. Ng, R. Howard, P. K. McGuire, and D. K. Jones, "Tract-specific anisotropy measurements in diffusion tensor imaging," *Psychiatry Res.*, vol. 146, no. 1, pp. 73–82, 2006, doi:10.1016/j.pscychresns.2005.11.002.

[93] D. C. Alexander, C. Pierpaoli, P. J. Basser, and J. C. Gee, "Spatial transformations of diffusion tensor magnetic resonance images," *IEEE Trans. Med. Imaging*, vol. 20, no. 11, pp. 1131–1139, 2001, doi:10.1109/42.963816.

[94] H. J. Park, M. Kubicki, M. E. Shenton, A. Guimond, R. W. McCarley, S. E. Maier, R. Kikinis, F. A. Jolesz, and C. F. Westin CF, "Spatial normalization of diffusion tensor MRI using multiple channels," *NeuroImage*, vol. 20, no. 4, pp. 1995–2009, 2003, doi:10.1016/j.neuroimage.2003.08.008.

[95] H. J. Park, C. F. Westin, M. Kubicki, S. E. Maier, M. Niznikiewicz, A. Baer, M. Frumin, R. Kikinis, F. A. Jolesz, R. W. McCarley, and M. E. Shenton, "White matter hemisphere asymmetries in healthy subjects and in schizophrenia: a diffusion tensor MRI study," *NeuroImage*, vol. 23, no. 1, pp. 213–223, 2004, doi:10.1016/j.neuroimage.2004.04.036.

[96] S. M. Smith, M. Jenkinson, H. Johansen-Berg, D. Rueckert, T. E. Nichols, C. E. Mackay, K. E. Watkins, O. Ciccarelli, M. Z. Cader, P. M. Matthews, and T. E. Behrens, "Tract-based spatial statistics: Voxelwise analysis of multi-subject diffusion data," *NeuroImage*, vol. 31, no. 4, pp. 1487–505, 2006, doi:10.1016/j.neuroimage.2006.02.024.

[97]   S. M. Smith, H. Johansen-Berg, M. Jenkinson, D. Rueckert, T. E. Nichols, K. L. Miller, M. D. Robson, D. K. Jones, J. C. Klein, A. J. Bartsch, and T. E. Behrens, "Acquisition and voxelwise analysis of multi-subject diffusion data with tract-based spatial statistics," *Nat. Protoc.*, vol. 2, no. 3, pp. 499–503, 2007, doi:10.1038/nprot.2007.45.

[98]   A. Mechelli, C. J. Price, K. J. Friston, and J. Ashburner, "Voxel-based morphometry of the human brain: methods and applications," *Curr. Med. Imaging Rev.* vol. 1, pp. 105–113, 2005, doi:10.2174/1573405054038726.

[99]   C. D. Good, J. Ashburner, and R. S. Frackowiak, "Computational neuroanatomy: New perspectives for neuroradiology," *Rev. Neurol.*, vol. 157, pp. 797–806, 2001.

[100]   C. D. Good, I. S. Johnsrude, J. Ashburner, R. N. Henson, K. J. Friston, and R. S. Frackowiak, "A voxel-based morphometric study of ageing in 465 normal adult human brains," *NeuroImage*, vol. 14, pp. 21–36, 2001, doi:10.1006/nimg.2001.0786.

[101]   J. Ashburner, and K. J. Friston, "Voxel-based morphometry – the methods," *NeuroImage*, vol. 11, pp. 805–821, 2000, doi:10.1006/nimg.2000.0582.

[102]   J. Ashburner, and K. Friston, "Multimodal image coregistration and partitioning – a unified framework," *NeuroImage*, vol. 6, no. 3, pp. 209–217, 1997, doi:10.1006/nimg.1997.0290.

[103]   J. Ashburner, and K. Friston, "Why voxel-based morphometry should be used?" *NeuroImage*, vol. 14, no. 6, pp. 1238–1243, Dec. 2001, doi:10.1006/nimg.2001.0961.

[104]   R. Peyron, M. Frot, F. Schneider, L. Garcia-Larrea, P. Mertens, F. G Barral, M. Sindou, B. Laurent, and F. Mauguiere, "Role of operculoinsular cortices in human pain processing: converging evidence from PET, fMRI, dipole modeling, and intracerebral recordings of evoked potentials," *NeuroImage*, vol. 17, no. 3, pp. 1336–1346, 2002, doi:10.1006/nimg.2002.1315.

[105]   P. A. Valdes-Sosa, R. Kotter, and K. J. Friston, "Introduction: Multimodal neuroimaging of brain connectivity," *Philos. Trans. R. Soc. Lond. Biol. Sci.*, vol. 360, no. 1457, pp. 865–867, 2005, doi:10.1098/rstb.2005.1655.

[106]   J. Ashburner, J. G. Csernansky, C. Davatzikos, N. C. Fox, G. B. Frisoni, and P. M. Thompson, "Computer-assisted imaging to assess brain structure in healthy and diseased brains," *Lancet Neurol.*, vol. 2, pp. 79–88, 2003, doi:10.1016/S1474-4422(03)00304-1.

[107]   S. R. Felber "Magnetic resonance in the differential diagnosis of dementia," *Neural. Transm.*, vol. 109, no. 7–8, pp. 1045–1051, 2002, doi:10.1007/s007020200088.

[108]   W. L. Au, J. R. Adams, A. Troiano, and A. J. Stoessl, "Neuroimaging in Parkinson's disease," *Neural. Transm. Suppl.*, vol. 70, pp. 241–248, 2006.

[109]   S. Kalra, and D. Arnold, "neuroimaging in amyotrophic lateral sclerosis," *Amyotroph. Lateral Scler. Motor Neuron. Disord.*, vol. 4, no. 4, pp. 243–248, 2003, doi:10.1080/14660820310011269.

[110]   P. Bartenstein, "PET in neuroscience: dopaminergic, GABA/benzodiazepine, and opiate system." *Nuklearmedizin*, vol. 43, no. 1, pp. 33–42, 2004.

[111] M. Filippi and M. A. Rocca, "Conventional MRI in multiple sclerosis," *NeuroImaging*, vol. 17, Suppl. 1, pp. 3S–9S, 2007.

[112] M. Rovaris, and M. Filippi, "Diffusion tensor MRI in multiple sclerosis," *NeuroImaging*, vol. 17, Suppl 1, pp. 27S–30S, 2007.

[113] N. De Stefano, and M. Filippi, "MR spectroscopy in multiple sclerosis," *J NeuroImaging*, vol. 17, Suppl. 1, pp. 31S–35S, 2007.

[114] M. A. Rocca and M. Filippi, "Functional MRI in multiple sclerosis," *NeuroImaging*, vol. 17, Suppl 1, pp. 36S–41S, 2007.

[115] M. Filippi and F. Agosta, "Magnetization transfer MRI in multiple sclerosis," *NeuroImaging*, vol. 17, Suppl. 1, pp. 22S–26S, 2007.

[116] C. Weiller, A. May, M. Sach, C. Buhmann, and M. Rijntjes, "Role of functional imaging in neurological disorders," *Magn. Reson. Imaging*, vol. 23, no. 6, pp. 840–850, 2006, doi:10.1002/jmri.20591.

[117] K. Kamada, M. Moller, M. Saguer, J. Kassubek, M. Kaltenhauser, H. Kober, M. Uberall, H. Lauffer, D. Wenzel, and J. Vieth, "Localization analysis of neuronal activities in benign rolandic epilepsy using magnetoencephalography," *Neurol. Sci.*, vol. 154, no. 2, pp. 164–172, 1998, doi:10.1016/S0022-510X(97)00226-8.

[118] F. Babiloni, D. Mattia, C. Babiloni, L. Astolfi, S. Salinari, A. Basilisco, P. M. Rossini, M. G. Marciani, and F. Cincotti, "Multimodal integration of EEG, MEG, and fMRI data for the solution of the NeuroImage puzzle," *Magn. Reson. Imaging*, vol. 22, no. 10, pp. 1471–1476, 2004, doi:10.1016/j.mri.2004.10.007.

[119] B. D. Gonsalves, I. Kahn, T. Curran, K. A. Norman, and A. D. Wagner, "Memory strength and repetition suppression: Multimodal imaging of medial temporal cortical contributions to recognition," *Neuron*, vol. 47, no. 5, pp. 751–761, 2005, doi:10.1016/j.neuron.2005.07.013.

[120] C. H. Im, H. K. Jung, and N. Fujimaki, "fMRI-constrained MEG source imaging and consideration of fMRI invisible sources," *Hum. Brain Mapp.*, vol. 26, no. 2, pp. 110–118, 2005, doi:10.1002/hbm.20143.

[121] P. Jannin, O. J. Fleig, E. Seigneuret, C. Grova, X. Morandi, and J. M. Scarabin, "A data fusion environment for multimodal and multi-informational neuronavigation," *Comput. Aided Surg.*, vol. 5, no. 1, pp. 1–10, 2000, doi:10.1002/(SICI)1097-0150(2000)5:1<1::AID-IGS1>3.0.CO;2-4.

[122] P. Jannin, X. Morandi, O. J. Fleig, E. Le Rumeur, P. Toulouse, B. Gibaud, and J. M. Scarabin, "Integration of sulcal and functional information for multimodal neuronavigation," *Neurosurgery*, vol. 96, no. 4, pp. 713–723, 2002.

[123] O. Ganslandt, R. Fahlbusch, H. Kober, J. Gralla, and C. Nimsky, "Use of magnetoencephalography and functional neuronavigation in planning and surgery of brain tumors," *Nervenarzt*, vol. 73, pp. 155–161, 2002.

[124] H. Hund, A. R. Rezai, E. Kronberg, J. Cappell, M. Zonenshayn, U. Ribary, P. J. Kelly, and R. Llinas, "Magnetoencephalographic mapping: Basic of a new functional risk profile in the selection of patients with cortical brain lesions," *Neurosurgery*, vol. 40, pp. 936–942, 1997, doi:10.1097/00006123-199705000-00011.

[125] O. Ganslandt, R. Steinmeier, H. Kober, J. Vieth, J. Kassubek, J. Romstock, C. Strauss and R. Fahlbusch, "Magnetic source imaging combined with image-guided frameless stereotaxy: A new method in surgery around the motor strip," *Neurosurgery*, vol. 41, pp. 621–627, 1997, doi:10.1097/00006123-199709000-00023.

[126] O. Ganslandt, M. Buchfelder, P. Hastreiter, P. Grummich, R. Fahlbusch, and C. Nimsky, "Magnetic source imaging supports clinical decision making in glioma patients." *Clin. Neurol. Neurosurg.*, vol. 107, pp. 20–26, 2004, doi:10.1016/j.clineuro.2004.02.027.

[127] P. Ossenblok, F. S. Leijten FS, J. de Munck, G. J. Huiskamp, F. Barkhof, and P. Boon, "Magnetic source imaging contributes to the presurgical identification of sensorimotor cortex in patients with frontal lobe epilepsy," *Clin. Neurophysiol.*, vol. 114, pp. 221–232, 2003, doi:10.1016/S1388-2457(02)00369-3.

[128] J. S. Ebersole, "Magnetoencephalography/magnetic source imaging in the assessment of patients with epilepsy," *Epilepsia*, vol. 38, Suppl 4, pp. S1–S5, 1997, doi:10.1111/j.1528-1157.1997.tb04533.x.

[129] M. J. Fischer, G. Scheler, and H. Stefan, "Utilization of magnetoencephalography results to obtain favourable outcomes in epilepsy surgery," *Brain*, vol. 128, pp. 153–157, 2005, doi:10.1093/brain/awh333.

[130] H. Stefan, C. Hummel, G. Scheler, A. Genow, K. Druschky, C. Tilz, M. Kaltenhauser, R. Hopfengartner, M. Buchfelder, and J. Romstock, "Magnetic brain source imaging of focal epileptic activity: a synopsis of 455 cases," *Brain*, vol. 126, pp. 2396–2405, 2003, doi:10.1093/brain/awg239.

[131] W. J. Marks Jr., "Utility of MEG in presurgical localization," *Epilepsy Curr.*, vol. 4, pp. 208–209, 2004, doi:10.1111/j.1535-7597.2004.04516.x.

[132] R. Firsching, I. Bondar, H. J. Heinze, H. Hinrichs, T. Hagner, J. Heinrich, and A. Belau, "Practicability of magnetoencephalography-guided neuronavigation," *Neurosurg. Rev.*, vol. 25, pp. 73–78, 2002, doi:10.1007/s101430100161.

[133] A. M. Dale, and E. Halgren, "Spatiotemporal mapping of brain activity by integration of multiple imaging modalities," *Curr. Opin. Neurobiol.*, vol. 11, pp. 202–208, 2001, doi:10.1016/S0959-4388(00)00197-5.

[134] B. A. Assaf, K. M. Karkar, K. D. Laxer, P. A. Garcia, E. J. Austin, N. M. Barbaro, and M. J. Aminoff, "Magnetoencephalography source localization and surgical outcome in temporal lobe epilepsy," *Clin. Neurophysiol.*, vol. 115, pp. 2066–2076, 2004, doi:10.1016/j.clinph.2004.04.020.

[135]   F. Duffner, D. Freudenstein, H. Schiffbauer, H. Preissl, R. Siekmann, N. Birbaumer, and E. H. Grote, "Combining MEG and MRI with neuronavigation for treatment of an epileptiform spike focus in the precentral region: a technical case report," *Surg. Neurol.*, vol. 59, pp. 40–45, 2003, doi:10.1016/S0090-3019(02)00972-2.

[136]   S. Knake, P. E. Grant, S. M. Stufflebeam, L. L. Wald, H. Shiraishi, F. Rosenow, D. L. Schomer, B. Fischl, A. M. Dale, and E. Halgren, "Aids to telemetry in the presurgical evaluation of epilepsy patients: MRI, MEG and other noninvasive imaging techniques," *Suppl. Clin. Neurophysiol.*, vol. 57, pp. 494–502, 2004.

[137]   K. Kamada, T. Todo, Y. Masutani, S. Aoki, K. Ino, A. Morita, and N. Saito, "Visualization of the frontotemporal language fibers by tractography combined with functional magnetic resonance imaging and magnetoencephalography," *Neurosurgery*, vol. 106, no. 1, pp. 90–98, 2007.

[138]   C. Ramon, J. Haueisen, T. Richards, and K. Maravilla "Multimodal imaging of somatosensory evoked cortical activity," *Neurol. Clin. Neurophysiol.*, vol. 96, 2004.

[139]   A. Stadlbauer, C. Nimsky, S. Gruber, E. Moser, T. Hammen, T. Engelhorn, M. Buchfelder, and O. Ganslandt, "Changes in fiber integrity, diffusivity, and metabolism of the pyramidal tract adjacent to gliomas: a quantitative diffusion tensor fiber tracking and MR spectroscopic imaging study," *AJNR Am. J. Neuroradiol.*, vol. 28, no. 3, pp. 462–469, 2007.

[140]   M. P. Richardson, K. J. Friston, S. M. Sisodiya, M. J. Koepp, JAshburner, S. L. Free, D. J. Brooks, and J. S. Duncan, "Cortical grey matter and benzodiazepine receptors in malformations of cortical development. A voxel-based comparison of structural and functional imaging data." *Brain*, vol. 120, pp. 1961–1973, 1997, doi:10.1093/brain/120.11.1961.

[141]   J. Kassubek, H. J. Huppertz, J. Spreer, and A. Schulze-Bonhage, "Detection and localization of focal cortical dysplasia by voxel-based 3-D MRI analysis," *Epilepsia*, vol. 43, pp. 596–602, 2002, doi:10.1046/j.1528-1157.2002.41401.x.

[142]   S. B. Antel, A. Bernasconi, N. Bernasconi, D. L. Collins, R. E. Kearney, R. Shinghal, and D. L. Arnold, "Computational models of MRI characteristics of focal cortical dysplasia improve lesion detection," *NeuroImage*, vol. 17, pp. 1755–1760, 2002, doi:10.1006/nimg.2002.1312.

[143]   H. J. Huppertz, C. Grimm, S. Fauser, J. Kassubek, I. Mader, A. Hochmuth, J. Spreer, and A. Schulze-Bonhage, "Enhanced visualization of blurred gray-white matter junctions in focal cortical dysplasia by voxel-based 3D MRI analysis," *Epilepsy Res.*, vol. 67, pp. 35–50, 2005, doi:10.1016/j.eplepsyres.2005.07.009.

[144]   J. Kassubek, F. D. Juengling, and A. Otte, "Improved interpretation of cerebral PET with respect to structural MRI changes using combined voxel-based statistical mapping," *Eur. J. Nucl. Med.*, vol. 27, pp. 880, 2000.

[145]   J. H. Cross, "Functional neuroimaging of malformations of cortical development," *Epileptic Disord*, vol. 5, Suppl 2, pp. S73–S80, 2003.

[146] R. Wiest, J. Kassubek, K. Schindler, T. J. Loher, C. Kiefer, L. Mariani, M. Wissmeyer, G. Schroth, J. Mathis, B. Weder, and F. D. Juengling, "Comparison of voxel-based 3-D MRI analysis and subtraction ictal SPECT coregistered to MRI in focal epilepsy," *Epilepsy Res.* vol. 65, pp. 125–133, 2005, doi:10.1016/j.eplepsyres.2005.05.002.

[147] F. D. Juengling, J. Kassubek, H. J. Huppertz, T. Krause, and T. Els, "Separating functional and structural damage in persistent vegetative state using combined voxel-based analysis of 3-D MRI and FDG-PET," *Neurol. Sci.*, vol. 228, no. 2, pp. 179–184, 2005, doi:10.1016/j.jns.2004.11.052.

[148] J. Daunizeau, C. Grova, G. Marrelec, J. Mattout, S. Jbabdi, M. Pelegrini-Issac, J. M. Lina, and H. Benali, "Symmetrical event-related EEG/fMRI information fusion in a variational Bayesian framework," *NeuroImage*, vol. 36, no. 1, pp. 69–87, 2007, doi:10.1016/j.neuroimage.2007.01.044.

[149] B. Fabio, B. Claudio, C. Filippo, R. Paolo Maria, B. Alessandra, A. Laura, and C. Febo, "Multimodal integration of EEG and functional magnetic resonance recordings," in *Proc Conf IEEE Eng. Med. Biol. Soc.*, vol. 7, pp. 5311–5314, 2004, doi:10.1109/IEMBS.2004.1404483.

[150] P. Ritter and A. Villringer, "Simultaneous EEG-Fmri," *Neurosci. Biobehav. Rev.*, vol. 30, no. 6, pp. 823–838, 2006, doi:10.1016/j.neubiorev.2006.06.008.

[151] F. S. Leijten, G. J. Huiskamp, I. Hilgersom, and A. C. Van Huffelen "High-resolution source imaging in mesiotemporal lobe epilepsy: a comparison between MEG and simultaneous EEG," *Clin. Neurophysiol.*, vol. 20, pp. 227–238, 2003, doi:10.1097/00004691-200307000-00001.

[152] J. Gotman, E. Kobayashi, A. P. Bagshaw, C. G. Benar, and F. Dubeau, "Combining EEG and fMRI: A multimodal tool for epilepsy research," *Magn. Reson. Imaging*, vol. 23, no. 6, pp. 906–920, 2006, doi:10.1002/jmri.20577.

[153] G. Lantz, L. Spinelli, R. G. Menendez, M. Seeck, and C. M. Michel, "Localization of distributed sources and comparison with functional MRI," *Epileptic Disord. (Special Issue)*, pp. 45–58, 2001.

[154] T. Hammen, M. Schwarz, M. Doelken, F. Kerling, T. Engelhorn, A. Stadlbauer, O. Ganslandt, C. Nimsky, A. Doerfler, and H. Stefan, "1H-NMR spectroscopy indicates severity markers in temporal lobe epilepsy: Correlations between metabolic alterations, seizures, and epileptic discharges in EEG," *Epilepsia*, vol. 48, no. 2, pp. 263–269, 2007, doi:10.1111/j.1528-1167.2006.00856.x.

Author Biography

Dr. rer. nat. Hans-Peter Müller is currently working as researcher at the Department of Neurology, University of Ulm, Germany. He formerly worked at Central Institute for Biomedical Engineering, University of Ulm, Germany and at Physikalisch-Technische Bundesanstalt, Berlin, Germany. He finished his Ph.D. in 1996 in magnetic resonance microscopy. Main areas of interest include software development for magnetic resonance imaging (diffusion tensor imaging as well as functional magnetic resonance imaging) and magnetoencephalography.

Prof. Dr. med. Jan Kassubek is currently working as senior clinician and researcher at the Department of Neurology, University of Ulm, Germany and has the position as the Vice Chairman. He finished his M.D. in 1997. He formerly worked as Medical Doctor/Researcher at the Division of Experimental Neuropsychiatry, Department of Neurology, University of Erlangen, Germany, then at the Center for Magnetic Resonance Research, University of Minnesota Medical School, Minneapolis, MN, USA, and at the Department of Neurology, University of Freiburg, Germany. Main areas of interest are clinical research on neuroimaging, in particular magnetic resonance imaging and magnetoencephalography, with a focus on neurodegenerative diseases.